$12.75

Anna et al...
Thanks so much
for visiting.
I hope your trip back
is safe and great...
lots of love,
Josie

Following the Bloom

Also by Douglas Whynott

A Unit of Water, A Unit of Time:
Joel White's Last Boat

Giant Bluefin

Following the Bloom

Across America with the Migratory Beekeepers

DOUGLAS WHYNOTT

JEREMY P. TARCHER/PENGUIN

a member of Penguin Group (USA) Inc.

New York

Most Tarcher/Penguin books are available at special quantity discounts for bulk purchase for sales promotions, premiums, fund-raising, and educational needs. Special books for book excerpts also can be created to fit specific needs. For details, write Penguin Group (USA) Inc. Special Markets, 375 Hudson Street, New York, NY 10014.

JEREMY P. TARCHER/PENGUIN
a member of
Penguin Group (USA) Inc.
375 Hudson Street
New York, NY 10014
www.penguin.com

First published in 1991 by Stackpole Books

First Jeremy P. Tarcher/Penguin Edition 2004

Library of Congress Cataloging-in-Publication Data

Whynott, Douglas, date.
Following the bloom : across America with the migratory beekeepers /
Douglas Whynott.
p. cm.
Originally published: Harrisburg, Pa. : Stackpole Books, c1991.
Includes bibliographical references (p.).
ISBN 1-58542-280-0 (alk. paper)
1. Migratory beekeepers—United States. 2. Bee culture—United States.
3. Honeybee—United States. 4. Africanized honeybee—United States.
I. Title.
SF524.5.W48 2004 2003061382
638'.1'0973—dc22

Printed in the United States of America
1 3 5 7 9 10 8 6 4 2

This book is printed on acid-free paper. ♾

Book design by Stephanie Huntwork

For Vernon and Marilyn Whynott

Acknowledgments

My thanks for the support of the Agricultural and Resource Economics Department at the University of Massachusetts—to Cleve Willis, Richard Rogers, Barry Field, Glenn Caffrey, Eileen Keegan, Ellen Scoville, Peggy Cialek.

My thanks for the support of those who helped, suggested, advised, or hosted—to Richard Todd, Mark Kramer, Sarah Provost, Bill Wishbow, Carol Steele, Michael Burke, Mary Wall and Wally Lynch, Tomas and Bojana Warchol, Richard and Laurie Norton, Louis Bourgeois, Bonnie Miller, Judy Umana, Lisa Butler, Kathy Whynott, Jim and Elinor Condon.

My thanks to Richard Parks, and to Sally Atwater and Jennifer Byrne.

My thanks to Jay Neugeborn, writer in residence at the University of Massachusetts, teacher, adviser, reader, friend.

And thanks to all of those beekeepers who answered my questions and let me ride along for a while, and especially to Andy Card, Junior.

Preface to the New Edition

I am pleased to preface this new Tarcher/Penguin edition of *Following the Bloom.* Not only is it a pleasure to see this book handsomely reproduced and reissued; I've also been afforded the pleasure of revisiting the past, of catching up with Andy Card again. I've had the chance to reinhabit a big dream once dreamt long ago.

One of the great things about this kind of nonfiction writing is that the characters continue to develop on their own, while you're away. Andy Card is a different sort of beekeeper than he was in 1985. ("Bee all you can bee," Andy says of recent efforts to diversify.) His sons are grown now. The little boy delighting in springing open a pop-top can in the original narrative is now a welder with an agricultural management degree. The toddler in a high chair is a college sophomore studying the same field. Both are strong young men ready to pick up the ball (or rather, the hive) and make a run for the honey.

Preface to the New Edition

Beekeeping has changed, as everything does, but beekeeping is the same too. Certain kinds of rejuvenation can occur. Mites have come, the Africanized bees have come, and beekeepers have adjusted. Honey prices have fluctuated, but production in the United States is roughly the same as it was when *Following the Bloom* was first published in 1991. There are approximately as many commercial beekeepers as before, though some have more hives now. There seem to be as many hobbyist beekeepers as before—perhaps even more.

In writing a new epilogue, I not only get to see some of the characters again, more fully developed; I also get to follow the bloom again, and see the landscape as a floral economy through the eyes of a bee-thinker. As Andy Card says, everyone should be able to do that, at least for a little while.

Andy has said of his depiction in this book, "I was a legend before I deserved to be one." Deserving or not, the legend continues, in this particular form.

By way of acknowledgment, I'd like to thank Ashley Shelby, the editor responsible for producing this new edition.

Cast of Characters

IN ORDER OF APPEARANCE

Andy Card, Jr., Merrimack Valley Apiaries, Billerica, Massachusetts. 4,000 beehives, used primarily for pollination, from Florida to New Jersey to Maine. Largest pollinating company in eastern United States.

Crystal Card, Billerica, Massachusetts. Andy Card's wife and queen-bee breeder.

Dale Thompson, Billerica, Massachusetts. Truck driver and beekeeper for Merrimack Valley Apiaries.

Jeff Kalmes, Billerica, Massachusetts. Truck driver and beekeeper for Merrimack Valley Apiaries.

Horace Bell, Deland, Florida. 12,000 hives in 1985, used primarily for honey production in Florida and the Dakotas. One of three top honey producers in the United States.

Luella Bell, Deland, Florida. A working beekeeper; Horace Bell's wife.

Nina Bell, Deland, Florida. Horace Bell's daughter, working in the business.

Gary Kelso, Deland, Florida. Nina Bell's husband, working in the business.

Wayne Knight, Deland, Florida. One of Horace Bell's managers.

Wayne Carter, Blue Hill, Maine. Beekeeper for Horace Bell and Andy Card.

Tom Charnock, Charnock Apiaries, West Palm Beach, Florida. 3,000 hives, honey and pollination, running the "triangle" from Florida to Maine to North Dakota.

Reggie Wilbanks, Wilbanks Apiaries, Claxton, Georgia. Bee breeder, in 1985 selling 15,000 packages of bees and 50,000 queen bees.

Glenn Gibson, Minco, Oklahoma. Retired commercial beekeeper, lobbyist for the American Honey Producers.

Jim Owens, Sr., Mays Landing, New Jersey. A bee broker in New Jersey blueberry region.

Jimmy Owens, Jr., Mays Landing, New Jersey. Working with his father.

Sandy Goddard, Charnock Apiaries. Tom Charnock's business manager and girlfriend.

Judy Carlson, Chief apiary inspector, North Dakota.

Fred Tiffany, Coleharbor, North Dakota. 1,600 hives, moved from California and Texas to North Dakota.

Joe Romance, Coleharbor, North Dakota. 1,800 hives, moved from California to North Dakota.

Chris English, Coleharbor, North Dakota. Artist, college teacher, wanderer, working for Joe Romance.

Everett Kehm, Elgin, Texas, to Minot, North Dakota. 30 years in migratory beekeeping.

Bill Hurd, Avon, Florida, to Bottineau, North Dakota. Second-generation migratory beekeeper.

LaVerne Prettyman, Avon, Florida, to Dunseith, North Dakota. Operation depopulated in North Dakota.

Roger Bracken, sharecropper, Devil's Lake, North Dakota.

Francis Andress, sharecropper, Hettinger, North Dakota.

Tom Emde, New Salem, North Dakota, and Apopka, Florida. 800 hives, honey production.

Glossary

African honeybee. Two of the races of honeybees *(Apis mellifera scutella* and *Apis mellifera adansonii)* imported from Africa to Brazil in 1957.

Africanized honeybee. The hybridized bee, sometimes called *Apis mellifera brasilia*, the progeny of twenty-six swarms of African bees that escaped from a test apiary in Brazil, interbred with European bees, and subsequently "Africanized" Latin America.

American foulbrood, AFB. A contagious disease of bee larvae caused by a spore-producing bacterium, *Bacillus larvae.* Once the most prominent problem in beekeeping.

Beebread. Pollen stored in the cells of combs.

Bee liner. One who tracks feral bees to their nests—usually in the hollows of trees. Feeding a bee, he observes its direction of flight, awaits its return, and calculates the distance to the hive.

Bee space. That distance, between combs or between frame and hive wall, that bees "respect," and don't fill with wax. Discovered by

Lorenzo Langstroth in 1851, the concept of bee space made possible beekeeping on a commercial scale.

Bobcat. A forklift, with four-wheel drive, a small fifth wheel in the rear, and a tilting mast—used for moving beehives on pallets. The Bobcat brought on a new era of commercial beekeeping—migratory beekeeping.

Brood. The developing stages of bees, including egg, larvae, and pupae.

Brood chamber. That portion of the beehive, either one or two boxes or chambers, where bees nest. The brood chamber is not usually harvested for honey, as are the supers, above the brood chamber.

Brood comb. Those combs in the section of the hive where brood is raised.

Brood food. Salival compounds manufactured by nurse bees and fed to larvae.

Cell. A single hexagonal unit in a comb, used to store honey or pollen or to raise brood.

Cluster. A mass of bees grouped into a tight unit. There is usually a cluster of bees at the center of the nest; bees form winter clusters to contain heat; bees form swarm clusters of branched chains and a quasihollow core during their process of migration.

Colony. A community of bees, containing a queen, thousands of workers (sterile females), and usually a few hundred drones (male bees).

Drone. The male bee. The drone has no stinger and does not gather nectar; the purpose of the drone is mating and, perhaps, colony morale.

Excluder. A fence, or screen, placed at the entrance of a hive or between the brood chamber and the super, used to confine the queen bee, which cannot pass through because of her broader thorax and larger abdomen.

Extractor. A machine that removes honey from the comb by centrifugal force.

Follower bee. One of the bees that read the dance of the scout bee.

Foundation. A sheet of wax, pressed in the configuration of honeycomb, used to form a base for honeycomb.

Frame. A wooden, rectangular unit, hung inside the hive, used to support, or "frame," honeycomb. The movable frame revolutionized beekeeping.

Geomenotactic. The ability to climb against gravity at a constant angle.

Haplodiploidy. A sex-determining trait. A queen lays unfertilized eggs that develop into males, and fertilized eggs that develop into sterile female workers. The evolution of haplodiploidy 150 million years ago predisposed development of insect castes and insect societies.

Hive. The structure that a bee colony lives in. *Hive* is also a verb: beekeepers can hive a swarm of bees.

Honey loan program. The federal program that guarantees beekeepers a certain price for their honey. Beekeepers take a loan on their crop and either repay all or part of it, or default and turn the honey over to the government.

Langstroth hive. The movable-frame hive, named after Lorenzo Langstroth, the minister and beekeeper who discovered bee space.

Larva. An undeveloped bee that has hatched from the egg but not yet entered the sealed pupal stage.

Nosema. A bee disease, caused by a protozoan parasite, that leads to dysentery.

Nuc. A miniature colony of bees used to start a hive. Small mating nucs are used in queen breeding. From *nucleus* colony.

Nurse bee. A female worker bee, usually among the youngest bees in the hive, which feeds brood food to the developing larvae and does the other tasks of raising brood.

Package. An artificial colony of bees shipped in a wire cage, usually with a queen, used to boost a hive or start a new hive.

Pheromone. A chemical signal or messenger used by insects. Bees have pheromones for mating, defense, nesting, and queen potency.

Pollination. The transfer of pollen, or male sexual matter, from the anthers of one flower to the stigma of another flower.

Pupa. The development stage that follows the larval stage, when the nascent bee is sealed in the cell and undergoes metamorphosis.

Queen. The ovulant female bee—created by differential and intense feeding, in a process called trophogenesis—at the center of the colony, that lays eggs, as many, in some cases, as several thousand per day.

Queen cell. The elaborated queen cup, built to completion, containing the pupal queen.

Queen cup. A manufactured wax cup used in the raising of a queen. Larvae are transferred, or grafted, into queen cups, which are deposited into queenless hives and then elaborated by nurses.

Royal jelly. One of the brood foods, and a particularly important one. Fed to all larvae for at least three days, but fed to a queen for the entire larval period, royal jelly is the compound that determines that a bee will become a queen.

Scout bee. The worker bee in the field force that seeks floral sources or nesting places, and reports its findings to others in the hive.

Smoker. A device, usually a can with spout and hand bellows, used to blow smoke at bees, so as to calm them and prevent them from stinging.

Split. A beehive made by dividing another hive. Usually, several frames are installed in a new box, and a queen is added. Among commercial beekeepers, splitting is the most common way of making hives.

Super. The box, or hive part, used by bees to store surplus honey.

Swarm. A migrating colony of bees. Swarming is a form of group reproduction or colony division, like a cell dividing.

Swarm getting. The archaic beekeeping practice of keeping bees in small chambers, forcing swarms, gathering these swarms, and taking their honey at the end of the season—killing the bees in the process. Swarm getting predated the Langstroth hive and the movable frame.

Syrup. Sugar solution fed to bees in place of honey. Among commercial beekeepers, syrup usually means high-fructose corn syrup.

Tracheal mite. *Acarapis woodi,* a microscopic mite that infests honeybees' breathing tubes. Discovered in the United States in 1984.

Trek swarm. Afrikaner term referring to a practice among some races of African bees of abandoning a hive and flying long distances, often sixty to eighty miles, usually in search of new floral sources.

Trophogenesis. The process of determining caste by differential feeding—i.e., creating a queen bee by feeding her royal jelly.

Tropo-olfactory sense. The ability to use antennae both to feel and to smell. This sense is of great use to bees in finding their way around the flower, and in following the dances of scout bees.

Varroa mite. *Varroa jacobsoni,* a parasitic mite, which can be seen with the unaided eye. Discovered in the United States in 1986, it attacks developing larvae.

Vibratory episode. The act of creating movement and sound to communicate the distance and direction of a floral source or nesting site.

Wagtail dance. The movement in which the scout bee communicates direction by geomenotaxis (moving at an angle against gravity) and distance by vibrating its abdomen at a telling frequency.

Worker. A sterile female, a sister bee, that raises young, makes wax and comb, gathers nectar, makes honey, defends the colony, makes queens, scouts locations, leads swarms, and communicates by means of symbolic language.

We're the last real cowboys, the last people moving livestock across the United States.

1

According to the U.S. Department of Agriculture, there are 212,000 beekeepers in the United States, and they own about four million hives of honeybees. About two thousand of these beekeepers are professionals, commercial beekeepers, with large colony counts and full-time crews. This 1 percent of the beekeeping population owns 50 percent—two million—of the hives.

Half of the commercial beekeepers, roughly a thousand of them, migrate with their bees. They follow the bloom, north in the summer, south in the winter. Some call them wizards—they can take one hive, and in a few weeks have four—and thus make something from nothing. They have staggering workloads at times, up all night moving bees. They are like the old-style cowboys who herded cattle across the open ranges, except they move their bees by trailer trucks over the interstate highways and country roads.

They are aggressive and shrewd businesspeople. One man moved thirteen thousand hives from Florida to South Dakota and in five weeks made a million pounds of honey. Another migratory beekeeper takes four thousand hives to Florida for the winter and then rents them in the spring to blueberry, apple, and cranberry growers for pollination. It's a run for the honey, American style, and there are nearly as many routes as there are beekeepers.

Commercial beekeepers make 90 percent of their income from honey, and produce about two hundred million pounds of honey a year. Beekeepers have been protected through honey support programs since 1950, but they are not protected because of the honey they produce.

The greater value of the beekeepers and their stock, by a hundredfold, is in pollination, the transfer of male sexual matter, or pollen, from one flower to another. Bees—wild bees, bumblebees, honeybees—pollinate fruit, vegetable, and seed crops that have a combined value estimated at $20 billion. Honeybees do about 80 percent of that pollination. Migratory commercial beekeepers provide virtually all of the honeybees for pollination on large farms.

In 1990 the Africanized bee closed in on the United States border. This potentially devastating migration of the so-called killer bees follows a difficult decade for migratory beekeepers, and no year was more difficult than 1985. A microscopic parasite—the honeybee tracheal mite—was discovered in the winter before the 1985 season. Quarantines were imposed, travel was restricted, entire outfits were "depopulated," or exterminated. The honey support program was nearly eliminated in Congress. Also in 1985 there was an appearance of Africanized bees in California, offspring of a colony that may have been transported on oil equipment moved from Central or South America. The hazards and casualties of pesticide use, of spray kills, continued. And as always there was the persistent public misconception about bees, and the entomophobic reaction to them. The occasional highway accidents brought such headlines as "Swarm of Angry Bees Attacks Cops." So as barbed wire halted the range cowboy, regulations threatened the entomological cowboy.

Questions arise. As the Africanized bee moves into the United States, will migratory beekeepers be able to travel? Can they afford the liability insurance? Will they cause the spread of killer bees or will they in fact control their movement? A broader question involves pollination. If hobbyist beekeepers quit in fear of *abejas furiosas*, and if commercial beekeepers are bankrupted, how will the food economy be affected? What about those billions of dollars' worth of bee-pollinated crops?

For millions of years the honeybee has been a migratory creature, moving ahead of ice ages, covering the land, expanding with the floral territory, following the bloom. The migratory beekeeper took a cue from the bee, and by horse and wagon, by boat, by train, by flatbed truck, by semi, he also followed the bloom. Both bee and beekeeper have a vision for floral economy. Ahead, twisting and turning down the road, are some of these beekeepers.

2

In Morocco, along the River Ziz, there is a race of honeybee that flies from oasis to oasis seeking nectar. Moroccan beekeepers wrap hives in jute sacks and move them on donkeys.

The races of honeybees that adapted to the African grasslands, and to the scattered, sudden rains, can move many miles in search of profitable blooms. The bees fill their stomachs with honey and fly in a cloudlike swarm led by scouts. African honey hunters set clay pipes in trees, hoping to hive these migrant swarms.

Swarms establish themselves in pottery jugs and wicker baskets, in cork cylinders, in empty wooden boxes, in holes in trees, in eaves and walls and chimneys: once, it's said, the bees made honey in a lion's head.

Egyptian beekeepers put beehives on barges and moved up the Nile, starting in the Sudan in late winter, ending in Cairo, the market, in summer. The pharaohs were embalmed in honey.

Russian beekeepers moved basket hives into the Caucasus mountains by oxcart, progressing with the season up the mountainsides.

Austrian beekeepers move bees into the buckwheat regions of Klagenfurst. Italian beekeepers move bees to orange and lemon plantations near the sea, and following this bloom, into the Calabrian mountains for wild thyme and sage. Sicilian beekeepers move into the Ragosa region for the honeyflow from carob trees. Algerian beekeepers move into the valleys of the Atlas mountains for citrus, eucalyptus, rosemary, lavender, and thyme. In Israel, after the orange bloom, hives are moved to the hills of Galilee for acacia, cactus, lavender, wild carrot, and thistle.

Yugoslavian beekeepers move hives into lime orchards and to the Istrian peninsula for sweet chestnut, and to the islands of the Dalmatian coast for rosemary. Slovenian beekeepers move up the Krawanken Alps. Greek beekeepers move basket hives to flows on clover, chestnut, wild sage, mountain savory, honeydew, and heath, to the pine forests of Macedonia, to the islands of Thakos for honeydew. On the islands of Ios, cork hives are carried to the fields of heather.

Spanish beekeepers move into orange and rosemary, and to the central plateau for thyme, lavender, and sainfoin. Australian beekeepers move cross-continent for the copious nectar flow from karri trees, for yields of six hundred pounds per hive.

In America, migratory beekeepers move from Oklahoma, Texas, and Mississippi to Iowa to pollinate apple and pear trees, strawberries and raspberries; they move to the long-grass prairie and the Red River Valley for sunflower, to the Big Horn Valley in Wyoming for alfalfa, to Wisconsin to pollinate apples, to Washington for snowberry and apple and cherry, to Oregon for clover and thistleberry and cherry and vetch, to Ohio for sunflower and apple, from Florida to New York to pollinate apples, from South Carolina to New Jersey to pollinate blueberry and cranberry, from Florida to Pennsylvania to pollinate fruit trees, to Maryland for apple and cherry and tulip poplar and peach, to West Virginia for apple, to Delaware for lima beans, to Virginia for sourwood and cucumber and melons, to Michigan for cherry and raspberry and blueberry, from Texas and Mississippi to Wisconsin for cranberry and cucumber and apple and cherry, to Colorado for alfalfa and sweet clover, to cantaloupe and tamarack in Arizona, to prickly pear in Utah, to clover and alfalfa in Wyoming, in California to almond orchards and to prune and pear and kiwi, to melon and sunflowers, to orange and to alfalfa, to star thistle and black sage and desert flowers in the Owens Valley, to Idaho for pollination of apple and pear and cherry and vegetable seed, and into West Texas for cotton and alfalfa, to the Rio Grande Valley for marigold, honeydew, squash, cucumber, cantaloupe, and salt cedar, and to the Texas high plains for kinnikinnik, to Florida for citrus and Brazilian pepper bush, mallaluca and palmetto and mango, to Georgia for gallberry and tulip poplar and tupelo and titi, to Massachusetts for apple, to Maine for blueberry, to Cape Cod for cranberry.

3

I've always liked bugs, or I should say, insects that fly. Things crawling under rocks, no. But a flowering bush in midsummer with a multitude of the aerodynamic wonders of the wasp family jigging all over it, yes: bumblebees, honeybees, hornets, yellow jackets, and those big, thin-waisted wasps with the long, poised abdomens—the finest imaginings of the collective wasp unconscious, if there is such a thing (but hadn't five orders of hymenoptera, and one of termites, a hundred and fifty million years ago, nearly simultaneously devised a social order that rivals human society in its complexity?). Those bugs, mad over the flowers, that was a mysterious natural beauty.

When I was a boy, an uncle kept a beehive in my grandfather's backyard. It was a lone hive, under an old, unpruned apple tree with branches long as a willow. The bees streamed through the branches, and in the summer they gave the apple tree, with its Magwitch hair, a diabolical look. I watched from the beach plum thicket, fifty feet away.

A decade later I experienced a different impression of honeybees. I was walking with my dog, and we were near home, coming down a hill, late in the afternoon. Down in a clearing, set in a pocket in a sumac grove, was a beehive. I was standing in the flight path to the hive. The bees were coming over my head, over my shoulders, floating down the hill, tracing long, elegant arcs to the landing board of the hive. The rays of the setting sun were slanting through the dusty air and the tree branches, and as the bees flew through this light, they became luminous points of fire, tinkerbells. The sun made the bees translucent and lit up the little drops of nectar they carried.

It was enough to get me down the hill to the hive. My dog took a sting on his nose immediately and ran away, but I crouched near the hive entrance and watched: landings, soft and silent; the crowd at the entrance, touching, testing, inspecting; and that smell of plants, raw nectar, blowing into the air.

I learned many things about bees later, from the neighbor who owned this hive and showed me how to open it up. We talked of starting a business. Since his last name was Joy, we'd call it Whynott & Joy Honey Company.

I sat by that hive until it was too dark to see, and it was still pleasant, with the smell of nectar and the slight press of the heat from inside. Everything about bees was so slight, and at the same time, so incredibly vast—one bee, flying two miles to a flower patch. This opposition was always there, and it was what the commercial beekeepers elaborated upon, the honeybee's grasp of numbers.

Bees inspire, they make people think and imagine and dream. With thoughts of bees, you look at the landscape with a new eye. I went to the library to read about them, and fortunately in my local library there were many old beekeeping books. Dr. Miller, with horse and wagon, making queens. Richard Taylor, telling of the joys of it. Pellett, on honey plants and nectar regions. Shaw, on commercial beekeeping. I read about Francis Huber, whose *New Observations of the Natural History of Bees* was published in 1792. Huber built a leaf hive, with glass sections that opened like a book. He was the first to write that queen bees mate outside the hive, that queens mate several times, that a queen bee could be reared from brood, and that worker bees lay unfertilized eggs. Huber went blind, partly from watching bees, but he continued his relentless observations, using a servant who lay on the ground under a hive and looked through a glass bottom, observing that female worker bees drag and expel male drones from the hive in the fall.

The next spring I started two hives. At a junk shop I bought a pile of equipment—old tools, veils, gloves, boxes. I bought nucleus, or starter, colonies and made more hives. The all-attentive beekeeper, I was forever opening the colonies up and inspecting them or manipulating them. The small colonies grew large, and one afternoon, with a hive opened wide, my smoker went out and I had my first encounter with stinging. For some reason, smoke deters bees from stinging, but with my smoker out, and the bees no longer deterred, they swarmed over me, digging into my shirt. Eighteen stings got through the fabric, and stunned, I dreamed frantic dreams that night.

By my third season I had eighteen hives in five locations. I harvested three hundred pounds of honey, and I went to the farmers' markets with honey in bottles, and with comb honey in boxes, and comb honey in jars. I was a rich man, and I gave presents of gold. I collected bee pollen and royal jelly and beeswax. And the following winter I considered going to a commercial beekeeping school in northern Alberta, one of the best beekeeping regions in the world. Though I changed my mind and went to graduate school instead, the floral economy of the northern Canadian plains continued to flicker in my imagination.

I became a member of two beekeeping clubs, and a representative on a committee to ban microencapsulated pesticides in Massachusetts. In a beekeepers' newsletter I read an advertisement for a summer job for a state bee inspector.

The job appealed to me for three reasons. One, I needed the money.

Two, I would meet beekeepers. Three, inspectors were not allowed to wear gloves, because gloves could become contaminated with the spores of American foulbrood, the disease that the inspectors were paid to eradicate. The most experienced beekeepers, I knew, worked without gloves. It meant taking stings on exposed fingers, but I wanted to take the risk, to learn. I wanted the touch, to work with bees on their own terms, to move with confidence in the bee yard. I had no use for the space-suit approach to beekeeping, and I didn't go for the sentimental approach, either, the beekeeper who would say that his bees didn't sting because they knew him, knew he had good vibes.

After I was hired, the chief inspector drove out from Boston for an orientation. He took me to the yard of a commercial beekeeper who had a case of American foulbrood, and showed me how to identify it. Then we opened hives for the rest of the afternoon, fifty, sixty, hives, in the hottest part of the day, with the bees roaring. No one wore gloves. I was not accustomed to being stung and I didn't have venom immunity. The chief, the former inspector, and the owner, they were used to stings. They didn't wince. They were cool, casually flicking out the stinger after the occasional sting. Bees hit my hands, and they crawled up my pant legs. I'd do a spin turn, facing off the unseen attacker.

The chief gave me a name tag, a smoker, a hive tool, a bucket, and a list of beekeepers in three counties in western Massachusetts. It was my job to knock on their doors and look into their beehives for sick bees, to issue treatment orders for the cases I found, to issue burn orders if the illness was too far advanced for treatment.

I happily took to the road. I inspected city hives, suburban hives, farm hives. One hive—its owner was in dental school and too busy to take care of it—had been piled up so high I needed a stepladder to take the top off. In one town where I found a pocket of American foulbrood the fire chief, a beekeeper, helped me locate all the beekeepers and clean things up. I moved through the Berkshires. I ate lunches in fields with extraordinary views.

I met James Vacelli, who had "lined" for bees that had settled into trees in the woods. He had a method of catching a bee in a baited trap, releasing it, timing its flight, watching the direction, and by formula, knowing how far to pace to the bee tree. He would cut the hive down, take out the honeycomb, and carry it home in buckets. One time, Vacelli said, he spent three days lining bees, only to arrive at his own backyard hive. In another town I met another beeliner who told me that in the late 1800s beeliners in western Massachusetts were paid bounties to exterminate colonies of the English honeybees, brought to Massachusetts in the seventeenth century. The English bee was then being replaced by the Italian bee, a breed more suitable for commercial beekeeping. English bees tended to swarm when the nectar flowed; Italian bees were builders, and stayed put.

I met a man who had kept bees for sixty years. He raised comb honey, and he had bred his stock for gentleness. Not only did he work without gloves, he didn't wear a veil or use smoke. As the days went by I got the touch, and I took fewer stings. I learned how to use smoke, how to trail it gently over the top bars of the hive and then wait for the smoke to take its calming effect. Then, one day in a town in the Berkshires, when the flower clusters on the locust trees were long and smoky white, I stopped at the home of two beekeepers, husband and wife, who kept a hive by their garden. They had wanted to inspect their colony for disease, but had been afraid to open it up. I took the hive apart, and laid out the combs, some full of clear locust honey. "Watch him, Ed," the wife said. "He doesn't wear gloves." She turned to me. "We read about people like you in books."

I had arrived as a bee inspector.

4

When the chief inspector gave me my training session, he told me I might be needed to help inspect the colonies of a commercial beekeeper in Billerica, a town north of Boston. Merrimack Valley Apiaries had four thousand hives, which were moved to Florida for the winter. The beekeeper had to have a certificate of inspection to take bees into Florida, so each year the chief and his crew spent two weeks going through the hives. Drawing on a slight hyperbolic streak, the chief said MVA bees were so mean, they flew at his windshield when he drove in the yard, and chased him down the street when he left.

The chief didn't call me to inspect the MVA bees, so I called Andy Card, Junior, who ran the company. His father owned MVA. Card told me about his route—down to Florida in the late fall for a honeyflow on Brazilian pepper bush, then to southernmost Florida for a plant called mallaluca. In March he moved into the orange groves for an orange flow, which lasted two weeks and paid for the trip south and got the colonies strong for the trip north. In April MVA moved into New Jersey, to pollinate blueberries. In May, into apple orchards in Massachusetts, New Hampshire, and Maine. Before the first of June the entire operation was pollinating in the Maine blueberry barrens. In June they pollinated cranberry in Plymouth and on Cape Cod. In July the operation poised in purple loosestrife swamps along the 128 beltway outside Boston.

Card invited me to visit him in the barrens in Maine. I had a friend who lived nearby, so I went, watched Card and two workers load 760 hives on two trailer trucks in two hours, and then listened to bee stories over beer and fried clams. I also went to Card's Greenwood Farm, which, as the bee flies, was by the Concord River. I was to go to Card's farm many times, in all seasons. On certain days, when the holding yard had a good colony count, you could hear the bees humming from Andy's house, a hundred yards away.

Andy Card was a man of action, planning on the run—where hives were going, where locations could be found, where crops could be found (crops that would make good honey), what was waiting for him down the road. I spent many hours riding with Andy Card. With his assistants I rode shotgun in the straight trucks and the tractor-trailers with loads of bees, went all night through muggy-smelling cranberry bogs and orchards in snowball bloom and the vast blueberry barrens. I went with MVA on the Nantucket ferry, to deliver a shipment of leased bees, 250 hives, to a cranberry bog. This was done at night, too, because after the sun goes down the bees return to their hives. In the hold of the ferry a few bees escaped the net and clustered on a ceiling fixture. "Don't worry," Andy's man told the captain. "They'll fly out to the flowers in the town when it gets light."

One January I rode to Florida with MVA. Andy and his family were in one semi. I rode in the other truck, an old Mack they called Bullwinkle, with Dale Thompson, Andy's driver and assistant beekeeper. The passenger seat in Bullwinkle was not fastened to the floor, and when we hit bumps the seat would bounce free. On the rougher roads—the Cross-Bronx Thruway was especially memorable—I had to suspend myself over the seat. And Dale wouldn't make pit stops. I thrashed around in the sleeper, where, according to normal trucker's protocol, I shouldn't even be. When we reached Deland, Florida, I was half-crazy, grinning at waitresses. Later that spring, Dale told me he'd really been impressed. No one had ever ridden that far in the passenger seat of Bullwinkle before.

Along the St. Johns River in Florida, Andy Card scouted for locations for his bees. It was all a man-to-man thing. At one hayfield, a family was haying. The farmer shut off his baler.

"The bee man," this farmer said. His daughter spun a string of cartwheels by her dad and the beekeeper. "I hear you wanna bring in some bees."

Talk about bees was often either understatement or overstatement. Better here to go low. "I'd like to bring in a few," Andy said.

"How many you got?"

"About a truckload."

The farmer liked the game, and he seemed to like the beekeeper. "How big a truck?"

Andy's eyes swept across the field, setting down at a far corner. "About forty foot."

"Don't know much about bees. I hear there's money in them, though."

"Not much money in them lately. I lost money these past two years with the freeze on orange. Looks like I might lose again this year, too. Got to get them up here and on the pollen quick."

Andy talked bees, and the charm of the beekeeper emerged. This was not lost on the farmer. "Well," he finally said, "let's look around. See what we can work out."

Andy Card walked over to his car and lifted out a cardboard box full of glass jars. "I got a little honey for you," Andy said.

Location secure.

Andy said I should meet the man who was, he said, the best beekeeper in the country, the man he had learned the most from. His name was Horace Bell. Horace was in the Super Bowl, as beekeepers go. Horace had just made a million pounds of honey on yellow clover in North Dakota.

We went to Horace's place on a Sunday afternoon. We were joined by Tom Charnock, the first beekeeper to migrate from Florida to Maine for blueberry pollination. Charnock flew his Cessna to Deland from West Palm Beach. He used the plane primarily to scout out nectar sources for his bee yards. We got into Andy Card's car and went for a ride, to Horace's bee yards. Horace talked of his plan to collect five thousand pounds of pollen and capture the market, his recent purchase of fifty thousand pounds of beeswax foundation. They all talked nectar flows.

"Mallaluca any good down there?" Horace asked Tom Charnock.

"Hell no," Charnock said.

"Maple gone by?" Horace asked.

"Nothing to go."

"Mistletoe on that tree there," Horace said. "That's in bloom. White pollen. Any mango crop down there?"

"Not a thing," Charnock said. He looked out the window. "That myrtle bush. That fixing to bloom?"

"That's fixing to bloom now," Horace said.

Andy laughed with a sound like a clap. "How do you know that stuff, Horace?"

"I got a pair of binoculars. I sit up there and watch the damn things."

We made a few stops to look at hives—fifty beehives on the foundation of a burned-out motel, fifty more hives in a clearing in a pine grove. Then we parked at the end of a dirt road, climbed over a sand pile, and walked down a utility company access road. Beehives lined the road. Each one had an inverted jar set into the cover, and each jar was part full of corn syrup. The three beekeepers walking between the two columns of hives had the took of a military review.

We walked by hundreds of hives and came to a clearing under some power lines. Here was an encampment of three hundred hives. A river of bees poured from this clearing and flowed down the power line cut. We walked into this river, and it parted for us.

"Live oak trees blooming," Horace said. "You can tell when the trees are not as thick with leaves." Horace snapped a twig off a low bush and squeezed the buds. "Huckleberries getting ready to bloom, too."

Then Horace knelt down in front of a hive. It was a weak hive, with only a few bees flying from it; this hive had been made up at Horace's plant a week or two before. A bee landed at the entrance, and Horace snatched it. He stripped the pollen from the bee's legs and popped one of the pellets into his mouth. He considered the taste for a moment, then turned to Tom Charnock. He handed him the other pellet. "Taste this," Horace said.

Charnock—wide shoulders, leather jacket, little yellow grain between his fingers—hesitated, shrugged, and then ate the tiny pellet. He, too, stood considering.

"That there is pine pollen," Horace Bell said. Andy Card shook his head. Horace was too much.

They talked bees, here, at other bee yards, along the road, at Horace's, on the way to the airport, walking across the runway to Charnock's plane, and while Charnock took to the air. And the conversations continued in my thoughts—through that winter and into the next season, when I sold two pianos, took a loan, gave my credit cards a workout, knocked on doors, begged rides and places to stay along the route of migratory beekeeping, and followed the bloom for a while.

5

The greatest legend in American beekeeping is about the discovery of bee space by a minister called Lorenzo Langstroth. The concept of bee space came to Langstroth while he was walking down a Philadelphia street in 1851. The inspiration was so profound, he wrote, that he had to refrain from shouting, "Eureka!"

Some beekeeping terms: A colony is that group of bees—one queen, a hundred to two hundred drones, twenty thousand to fifty thousand workers—that form a single, organic unit. A hive is that structure that the colony lives in, but *hive* is an elastic term: it can refer to the empty box or to the box and the bees in it. *Hive* and *colony* can be synonymous. *Hive* can also be used as a verb: A beekeeper can hive a colony of bees.

A colony of bees can migrate. When it does, leaving its hive, the colony is called a swarm. This word can also be a verb: Bees swarm. Swarming is a form of group reproduction, of colony fission—like a cell dividing, one colony becomes two. This is one of honeybees' two forms of reproduction—individual, by queen and egg; and group, by colony fission.

There are many ways of describing the hive: good and bad, weak and strong, sick and healthy, queenless and queenright, light and heavy (the amount of honey inside), deadouts and junk hives, nasty hives and gentle hives (stinging disposition), to name a few. Before Langstroth, and the Langstroth hive, there were log hives, gum hives, box hives, skep hives, and many others.

When bees build hives in hollow trees or boxes or any other cavity, they attach combs to the walls and ceilings. Beekeepers of old opened the hives, sometimes killing the bees with brimstone smoke, cut out the comb, and pressed the honey out. They tried to leave the nest intact, and in the spring the bees started over, building new comb.

Bees, by nature, build their combs with a space between them. This space is about three-eighths of an inch. At the time Lorenzo Langstroth apprehended this fact, beekeepers were setting rails, pieces of wood, across the top of a hive. The bees attached the comb to these rails, as they would to the ceiling of any cavity. During the harvest the beekeeper cut the comb from the sides and lifted out the rails. What Langstroth saw, that day in 1851, was that he could suspend wooden frames in the hive, three-eighths of an inch from the walls, and the bees would leave the spaces between frame and hive wall free of wax. They would respect "bee space."

The Langstroth movable-frame hive brought on an era of commercial beekeeping, by increasing the ways a beekeeper could manipulate a hive. The Langstroth hive is made of a bottom board, a hive body (a box for the nest), one or more honey supers (boxes where the bees store their surplus honey), and a cover. That is the external unit. Suspended in the hive bodies are nine or ten frames, hanging like folders in a file cabinet.

Langstroth hives can be divided, or split. A beekeeper can lift five frames from the hive, set them inside another hive body, and—if there are eggs from which the nurse bees can make a queen—one colony soon becomes two. Splitting is, actually, a form of artificial colony fission.

As the technology for rearing queens and feeding developed, and bee wizardry increased, the numbers within beekeeping operations became greater. A thousand colonies could become four thousand. And in a bad year, the beekeeper could cut back, from four thousand to one thousand, holding the equipment for better times.

Many inventions followed Langstroth's discovery of bee space. There was "foundation," a sheet of wax pressed into the basic hexagonal configuration of honeycomb cells. These cells were five to the linear inch—worker bee cells, for the bees that gather honey, and not the four-to-the-inch drone cells. The "excluder" was a grid or fence that worker bees could pass through, but not the larger queen—it kept her from laying eggs in the honey supers. With a smoker—a can with a chimney lid and squeeze bellows—a beekeeper could direct a stream of smoke along the tops of the frames, thus leaving behind such smoking things as cigars, shovelfuls of coals, and burning, rotten wood. The honey extractor used centrifugal force to spin the honey out of the movable frames. The first extractors were cranked by hand. Now some of the commercial operations have several sixty-frame extractors going simultaneously, with honey flowing into five-thousand-gallon settling tanks, then into ten-thousand-gallon holding tanks, and then into fifty-five-gallon drums.

Beekeepers have been moving hives since earliest times. With the Langstroth hive it became easier, and many beekeepers experimented. In *History of American Beekeeping* (1938), Frank C. Pellett tells the story of a honey dealer from Chicago by the name of Perrine who in the 1870s floated an apiary on the Mississippi. He bought a steamboat, a barge, a thousand colonies of bees, and hired fifteen men. They stacked the hives on the barge and set out from New Orleans in the spring. Soon, however, the steamboat broke down, and after repairs to the ferry there were problems with the barge. The beekeepers loaded the hives on the steamboat and continued upriver. No one in the crew knew the nectar flow sequences along the Mississippi, so honey crops were hit-or-miss. At first they anchored close to shore and let the bees fly across the water, but the bees got lost and got chilled in the cold winds, so the crew carried the hives from the boat into the fields each time they stopped. This didn't work out well, either, and they ultimately abandoned the hives in a field in Minnesota. Perrine concluded that floating apiaries weren't profitable, that the train might be the way to go. It was, and it wasn't.

J. S. Harbison was the first beekeeper to move colonies by train into the Sierra Nevada foothills for a wildflower flow, and the first beekeeper to send railcars of honey back east. In the 1860s he had transported the first honeybees to California. He started out in Pennsylvania, traveled to New York, took a ship south and crossed Panama overland, took a ship to San Francisco, and then sailed up the Sacramento River. After the turn of the century, when beekeeping had really caught on in California, a beekeeper called Migratory Graham in one season moved 161 railcars of bees on a cycle from almond groves to orange groves to alfalfa and clover and to mountain wildflower. Nephi Miller, who was once the largest beekeeper in the world, with thirty-two thousand colonies, in 1907 moved bees to California for the winter by train. Miller was among the first to make that migration.

The railroad workers, however, did not welcome these beekeepers and their cargo. A car of bees sometimes got moved to a side track and forgotten, with bees dying and wax melting. "Cooking," this is called. Hives unloaded in the railroad yard were moved by horse and wagon. Horses got stung, both those harnessed to the wagon and those in the fields along the way.

When flatbed trucks came into use in the 1920s, migratory beekeeping entered a new era. Beekeepers could load hives on a truck and drive to good nectar flows. There were rough trips along rutted country roads, and there were flat tires, and the beekeeper often had to change the tire under a load of raging beehives.

When evidence came that bee pollination improved crop yields, farmers began to lease beehives. Around 1910 Richard Barclay, a beekeeper from New Jersey, developed an extensive pollination business

Putting on honey supers in western New York. —GLENN CARD

with apple orchardists and blueberry growers. He moved his bees by horse and wagon. Pollination enterprises developed in California at the same time.

Larger trucks came along, larger loads of hives went down the road. Beekeepers put three hundred, four hundred, and sometimes nearly five hundred hives on a semi. The problem was labor. It took a crew all night to load that many hives, each weighing a hundred pounds or more. But in the midsixties beekeepers in North Dakota began experimenting with the Bobcat forklifts that had been developed for farm use, machines with sharp turning radii and tilting masts. They developed a hardwood pallet that could hold four to six hives, and lifted the entire unit on the truck. It worked wonderfully, and the beekeeping industry changed in a way it had not changed since Langstroth discovered bee space.

Forklifts were for a new generation of beekeepers. The older beekeepers resisted this level of mechanization, which was expensive, especially if one bought a semi to go with the forklift. Mechanization also meant a loss of hand contact with beehives. A good beekeeper can lift a hive and know from the feel how the colony is faring. The old boys didn't want to lose this skill. But the advantages were too great. One beekeeper could load 380 hives on pallets in two hours, by himself, and be off for Florida or Texas or North Dakota. During the 1960s and into the 1970s, when the interstate highways were completed, migratory beekeeping underwent a growth spurt.

A separate problem involved pesticide use. The beekeeper's stock, of course, is just as susceptible to pesticides as the next insect, and the beekeeper is often fighting against the wind, trying to keep his colonies strong as the fields around his hives are sprayed. There was a federal indemnity program for a time, but that was terminated. But in a twist of irony, the use of pesticides helped, because pesticides and herbicides eliminated many indigenous pollinators. The commercial beekeeper, and especially the migratory beekeeper, became that much more valuable to the agricultural economy.

Migratory beekeeping reached a peak during the early 1980s. In 1984 and 1985, however, the honeybee tracheal mite appeared, first in Texas and then in Florida. The tracheal mite had been spreading from Europe throughout the world and by 1984 was present in hives in Mexico. In 1984 the USDA Animal and Plant Health Inspection Service surveyed a fifty-mile strip along the Mexican border and found tracheal mites in Wasleco, Texas, in the outfit of Waylon Chandler, a commercial beekeeper who leased hives for pollination and made honey crops along the Rio Grande Valley. Since it was thought he was the only beekeeper in the United States with tracheal mites, his entire operation of three thousand hives was "depopulated." (Chandler couldn't believe the government would do that to him: he was a World War Two combat veteran. "They done gassed my hives," he said.)

Soon the tracheal mite was found in Florida, in hives that had been to New York, North Dakota, South Dakota, and a number of other states. Panic spread. More depopulations followed, and personal tragedies ensued: a number of beekeepers were ruined, the chief inspector in Florida died of a heart attack, and the official in charge of the tracheal mite program committed suicide after his early-retirement party. Georgia banned the transport of bees, and so did California, in order to protect the bee-breeding industry. In December 1984, with the number of tracheal mite finds increasing, the state of Florida was placed under federal quarantine. No beekeepers could leave the state, at least not with their bees.

Federal and state agencies were using the occasion to test their inspection programs. Officials wanted to know how well they could survey for mites and how well they could contain the infestation. The tracheal mite, some believed, was just a warmup for the arrival of the more harmful varroa mite, which would in turn be followed by the Africanized honeybee, the killer bee.

When news of the tracheal mite arrived, the migratory beekeepers were all trying to second-guess the agencies and inspectors. Andy Card hedged his bets. He sent fifteen hundred hives to Florida, about half of his colony count then, and was looking for other locations. Horace Bell suggested South Carolina, so Card drove there in his Mack truck. He scouted out the swamplands along the Savannah River for good pollen and nectar plants. Driving along, about sixty miles upriver from Savannah near a town called Millbury City, Card spotted a farmer who was trying to push a stray cow off the road. Andy got out of his truck and helped. "I see you're a farmer," he said. "I'm a farmer, too."

The farmer told Andy he looked like a trucker. Andy told him to read the side of the truck—he was in beekeeping—and he was looking for a place to set some bees. He talked bees, and beekeeping.

After a while the farmer said he had five thousand acres, and they might be able to find a spot for some bees. Andy said he had something for him. He went to his truck and got a case of honey.

6

In mid-March I drove south. I met Andy Card in Claxton, Georgia, and followed him through Statesboro and up to Sylvania, Georgia, a town close to the Savannah River. Andy's bees were all on the South Carolina side of the Savannah.

Andy drove by a stubbled cornfield, by a catfish pond, by longleaf pines with cones the size of pineapples. There were cones under the trees, in thick clusters that looked like shadows. He parked on the lawn between two trucks, an International flatbed and a Mack semi. Both had stencils on the door—Mernmack Valley Apiaries: Billerica, Massachusetts & Deland, Florida. On the bed of the International was a two-thousand-gallon tank, chained down, containing high-fructose corn syrup—feed for bees. On the deep-green door of the semi was an arched yellow stripe that looked like a raised eyebrow.

It had been a good day. The prospects for working his hives up to field strength looked promising. He was happy to be showing me around.

There was determination in his voice as he walked by Jeff Kalmes and Dale Thompson, his crew, telling them they'd be feeding, first thing tomorrow. He brushed his hand over Wesley's hair and stepped up into the trailer, where Crystal was getting dinner together.

Jeff was grilling hamburgers. Dale was sitting in a lawn chair, musing over a Michelob. He held out a can for Wesley. Wes was four and had just discovered the secret of the pop top, the pleasure of the spray. Wes popped and Dale took the pleasure from there. Dale had been working for MVA for eight years, since tenth grade, when he was part of the night crews that hand-loaded bees for the runs to Maine. Now Dale was a year-round beekeeper and the primary MVA driver. He was a veteran of many Florida-to-New-England bee runs.

"This could be it," Dale said, "the time of the fall of the bee business. Everything's locked up in quarantine. People can't move out of Florida. If they can't get out of Florida, they can't pay on their loans. That means foreclosure. So what do they do?"

"Sneak out," Jeff Kalmus said. He didn't seem all that concerned. Jeff also had worked for MVA since tenth grade. Jeff was more of the pure beekeeper than Dale, and he had refused to learn to drive a semi. Both of them knew bees, could work bees, and could talk bees; they looked alike and, with their Boston accents, sounded alike; other beekeepers called them Chip and Dale. Both had received offers to work for other companies.

But Jeff wasn't interested in talking bees then. "I caught a bass here today," he said, nodding in the direction of the pond, "waiting for Andy. Two pounds, threw it back. That pond is full of catfish. The guy who owns it feeds them corn every night. He taps on a pipe and they come right up to him."

Jeff looked at the syrup tank and said it had a slow leak. Syrup was dripping out onto the ground. But they weren't telling Andy until tomorrow. Andy would get upset.

The sky darkened, the grill glowed. And then, there was Andy, in the lighted doorway, arms flexed and out to his side, a rolled-up paper in one hand. "Did you see this?" Andy said. "This contract? They want four hundred hives, and it says here that every hive has to have a hundred twenty bees a minute flying. No way there's gonna be a hundred twenty bees a minute coming from a hive. Not in April."

"Maybe in July," Jeff said. "It would take a hive with three deep boxes to have a hundred and twenty bees a minute flying. In July."

"Unbelievable."

"Looks like we're not going there," Dale said.

"We're going, but not with a hundred twenty bees a minute. What do you think? Is this a way of getting out of the contract? They're gonna get out there with a stopwatch and time these bees, and if we don't have a hundred twenty bees a minute they're not gonna pay us, you think that's it?"

"They're gonna check you out, Andy," Dale said.

Jeff took the hamburgers inside. Crystal Card had set out potatoes, green beans, bread. But Andy paced.

"Sit down, Andy," Crystal said, with a look that said to give it a break. Crystal had her work cut out for her. They had been in Georgia for two days: a husband, two boys, two workers, all in a sixty-foot trailer with worn rugs and overused furniture and stubborn plumbing. But the location did have privacy and pleasant surroundings.

Crystal was the MVA secretary and often their dispatcher. She, too, had helped load beehives on semis. She had raised queen bees in Florida. She remembered a favorite breeder queen, named Rose. "Rose had good characteristics," Crystal said. "Kept an organized hive. She was gentle. Her bees made good brood patterns. The color of her brood wasn't all that good, kind of dark, but I don't really care about color."

Glenn Card, six months old, was foraging in the carpet. Crystal got up and fastened him into a high chair. With dinner, Andy relaxed. Earlier that day he had seen some healthy colonies at an apiary owned by Reggie Wilbanks, a beekeeper who raised queens and sold "package" bees, or starter colonies. And good bees made for good moods.

"Hey, Wes," Andy said. "What were we doing today?"

Wes said he was putting a stick into the ground; that was his job.

"What was Reggie doing with those hives? What size were they?"

Wesley held his hands a few inches apart. "Little hives."

"That's right. What was Reggie doing with them?"

"Holding the rectangles," Wesley said.

"That's pretty close. Those were frames. And what was on them?"

"Queens," Wesley said.

Andy tilted back and laughed. Wesley blushed and hid his face in a window curtain.

I left soon. Andy followed me outside. Breakfast would be at seven. There was no phone in the trailer, so he was driving into Sylvania to call some other beekeepers, and his father, to talk about this hundred-twenty-bees-a-minute clause.

He wouldn't sign the contract, after all, and by the next day Andy seemed to have forgotten about it. There were more pressing matters, like corn syrup and pollen.

7

In a 1979 book called *Bumblebee Economics*, Bernd Heinrich used a vivid term to describe a dramatic evolutionary trend that took place about a hundred million years ago. The term is the angiosperm explosion, and it refers to the ascendancy of flowering plants that produce enclosed seeds (Neo-Latin, *angi-*, enclosed, and Greek, *sperma*, seed). The angiosperms are distinguished from plants with naked seeds, gymnosperms, which include the conifers (commonly called evergreens) and their relatives. The gymnosperms are the older group, and during most of the Mesozoic era (230 million to 65 million years ago) they dominated the landscape. At the end of the Mesozoic era was an interval of rapid evolution, undoubtedly brought on by major climatic and geologic changes. It was at this time that the angiosperms spread rapidly across the globe while the gymnosperms retreated to the cold temperate and polar zones where we find them today.

The angiosperms seemed to have two important advantages over other plants: a diversity of life types, such as trees, shrubs, grasses, and herbaceous plants, which allowed them to adapt to a variety of environments; and a seed enclosed by a protective ovary. Encased in an edible fruit or nut, or locked within a clinging burr, or attached to a feathery, windblown structure, the seed of an angiosperm was equipped to travel far from the parent plant and colonize new territory. The naked seeds of the gymnosperms fell at the foot of the parent plant, and there they stayed.

Many of the angiosperms, such as grasses, were pollinated by the wind. But for others, reproduction occurred when passing insects happened to transfer pollen from the anther of one flower to the stigma of another. Those plants that had more insects on or near their flowers were better pollinated than others. Better pollination meant a greater number of seeds produced, which meant higher numbers of those plants. Within certain groups of ancestral angiosperms, those that could attract insects flourished, while those that didn't lost ground—literally. In evolutionary terms, the use of insects for pollination was a successful strategy; it allowed certain plants to reproduce in larger numbers over a longer period of time than others whose flowers were less appealing to insects.

But for us, looking back on the angiosperm explosion, it's almost as though one group of plants set out to solve a business problem: how to market and sell enough pollen to ensure corporate growth and survival. Pollination could be a hit-or-miss proposition: pollen might reach the right market . . . or it might not. The plants' new strategy made use of the flower, an advertisement loud enough to attract customers' attention, and nectar, a premium sweet enough to lure them but in rations small enough to require many visits to many flowers. The pollen itself, the real product, had to be sticky enough to adhere to the shopper. It was stored in pollen chambers that grew from fingerlike anthers, sure to touch anything stopping by for dinner.

Inestimable variations on the strategy arose. The insects liked some colors better than others, so flower color was one way to attract the customer. Many plants advertised with odor as well as color. In some communities, plants competed for time slots: if in a certain bog there was an opening at ten A.M., a time when the nectar flow was low, then a plant was likely to exploit that time, to fill that market niche. It made good economic sense.

But the evolution of plants is only half the story. While plants were competing with each other to get the insects' attention, the insects were evolving into better shoppers. Insects that had wings could glide quickly and easily from flower to flower. Formerly carnivorous wasps began to feed on pollen, which was easier to get and, gram for gram, provided three times as much protein as insect flesh.

One hundred and fifty million years ago, the evolution of primitive wasps led to the appearance of haplodiploidy, whereby unfertilized eggs develop into males and fertilized eggs develop into females. This arrangement enabled insects to form advanced societies consisting of a single fertile female and a specialized caste of sterile female workers nursing the colony's larvae. Natural selection favored this kind of altruism, with specialized individuals making a kind of sacrifice for the good of the colony—a sterile worker giving up reproductive rights, a guard giving up its life.

With such advantages, wasp societies radiated over the planet, altering plant life and contributing to the rise of insect-pollinated plants. It took some twenty million years for insects and plants to evolve into the smoothly functioning system we observe today, but in the scheme of planetary evolution that is short stuff.

Bees arose from wasps. Specializing on pollen, adapting to a strictly vegetarian diet, they speciated (twenty thousand species exist today) and adapted. Fossil records for bees are scant, but sometime between fifty million and twenty million years ago, the society of *Apis mellifera* (the honey-carrying bee) had evolved into its present form.

By then *Apis mellifera* could make wax, and fashion it into hexagonal combs, testing the thickness of the wax walls by, perhaps, sound pulses. These bees had learned bee space. They could read the angle of the sun and orient by it. They had tropo-olfactory sense—mapping by smell—the ability both to feel and to smell with their antennae, which helped them navigate around the flower and its nectar guideposts. In the dark hive tropo-olfactory sense helped them read maps drawn by other bees just in from the field. Honeybees had developed a complex array of glandular secretions—brood food to induce rapid larval growth; royal jelly to create queens; secretions to make honey from raw nectar, to give off alarms, to sexually attract drones, to mark nest sites, to inhibit the creation of queens and the development of a worker's ovaries. *Apis mellifera* developed a language for communicating flight goals.

The honeybee also developed the ability to cluster. The cluster is a model of the earth itself—a sphere of bees with a crust and a hot core. Once honeybees could cluster in an enclosed space, they could move from the tropical to the temperate zone. The development of the cluster also made migration possible. A colony of bees, flying like a cloud, could land on a tree limb and cluster until they figured out where to nest.

When the fall had come and the nectar sources had dried up, a colony of bees could amass in the hive, forming a dense unit with an insulating shell of bees slowly moving their wing muscles to generate heat. Inside the cluster, at the core, even in freezing weather, provided there was enough honey to make the heat, the bees could raise core temperatures to the eighties and nineties, warm enough to raise brood.

Raising brood—usually two weeks after the winter solstice, as the days were steadily lengthening—they were anticipating the first flowerings of spring, and on the warm days at the end of the winter, the cluster would break up, lose its form, and spread through the hive. Some bees would take to the air. But the nest always has something of the imprint of the cluster, the spherical shape, where the queen lays eggs and the brood develop into mature honeybees. As the days warm into spring, and the first pollens emerge, the cluster disperses from the hive into the air, casting into the circumferent fields.

If the honeybee cluster is a good metaphor for the earth, then the dispersion of the cluster during warm weather, a dispersion in unison with the opening of the flowers, is a good metaphor for the angiosperm explosion, origin of the cluster.

Some beekeepers call it busting, this springtime thing.

8

It was a hot morning, into the seventies before nine, when Andy Card and I rode from Sylvania across the Savannah River to South Carolina. The apiaries we visited were in the town of Milbury City, about fifty miles upriver from the city of Savannah. Andy had placed seven hundred hives in two clearings on a sixty-five-hundred-acre ranch. Deer were bold there. They grazed in the caretaker's garden, and looked fearlessly over their shoulders when she banged pots and yelled at them from her trailer.

The maple trees had already bloomed and the florets had fallen off the branches. The trees, just turning color, looked like mists of green. Other trees were coming along, the wild plums and cherries, the sweet gums and black gums and the redbuds.

We rode by the trailers and horse paddock along a dirt road through a stand of pine trees with trunks charred by a recent burnoff. There was a plowed strip of newly cleared land and a dammed-up swamp with bleached trees and green water. As we got close to the bee yard, bees crossed in front of us and flew alongside the car. When we stopped, bees flew at the windshield and bounced against the windows.

"They wouldn't be doing that," Andy said, "if there was a good carbohydrate source in the field, if there was a honeyflow."

Feeding had already begun. Jeff was at the spigot, leaning back when the bees, looking for corn syrup, tried to fly in his mouth. Dale was stacking the filled jars in a bin that he could move to the hives with the forklift. The jars were then set into holes on the hive covers, and since the jar lids had little nail holes in them, a shower of syrup would rain down into the hive. The bees would lick the syrup off combs and stick their tongues in the nail holes, and take to the air to find the source of this gift from above. They would find the steel tank of syrup and the two men, whom they would buzz and land on but not sting—this was a food alert and not a defense alert—and they would find the trail of syrup that came from the leak, lining up like cows at a trough, wing to wing, sucking corn syrup to deliver to their hives. This hive stimulation meant progressively more bees in the air as long as the feeding went on, so Dale and Jeff would move the truck every now and then to give the bees the slip, leaving a swirl above an empty parking place.

"Let's take a look at these bees," Andy said, and we took a stroll through the apiary. The seven hundred hives were arranged like miniature housing tracts, with the rows of pallets and the lanes between them. Each pallet had six hives, three entrances on each side, with the pallets placed so that the bees came into the long lanes as if coming out of garages into the street.

The hive entrances faced in such a way that Jeff or Dale could drive the forklift up on the blind side. This way the guard bees wouldn't see the driver and pour out at him. And when the hives were loaded on the truck, the finished load would have sidewalls facing out rather than hive entrances, which made for many fewer stings, and an easier job of tying down the load.

We walked in front of the entrances, through the flight paths. To avoid collision with these huge, careening man-objects, the bees swerved and swept by us in fast arcs. The sounds were like highway sounds, a light and generalized overtone mingled with close fast-driver sounds. One bee caught in my hair, zizzed like a car slipped out of gear, got free, and flew on. This bee traffic creates a wonderful sensation. It's in this stream of zip and cacophony that beekeepers get inspired and do their best thinking.

We didn't put on gloves or veils. Considering all the bees we passed and all the bees that touched us and all the bees that followed us to check us out and give warning and nudge at us, it was remarkable that I only took one sting on the shoulder. It had been a while since I'd been stung. It hurt. As is the way of the sting, the first sensation, the prick, is followed by the burn of the venom and then the swelling, which can close an eye or make it impossible to remove a ring from a finger. The bee's barbed stinger, with venom pump, imbeds in the elastic human skin, separates itself from the bee, and keeps pumping. The experienced beekeeper flicks the stinger out with a fingernail, and since he has built up a resistance to bee venom, fifteen minutes after a sting he might not know where he took it. But not me, that day.

Andy stopped, took count, pulled jar lids to look through the hole in the cover, and sometimes pulled a cover up. He assessed the landing boards. "Look at those hives," Andy said. "Look at them trucking that pollen in. This is a better place than Florida right now. A better place for population buildup. Maybe a better place, period."

He passed by hives with full jars of syrup and then stopped at one with no jars, only empty lids on top, and lifted a lid. Bees shot out of the hole. Andy slapped the lid down and pulled a bee from his hair. One bee made for my face.

"See the difference between the hives that have feed and these? You got thirty, forty bees per minute more over there."

We stood in the swirl. "When I first went south I thought, well, you just get them out of the cold weather and you'll lower the honey consumption.

That isn't what happens. You're really going south for the protein buildup. Even if you get the occasional freezes down here, you still get the early pollen flow.

"The migration is for raising bees. To raise bees and get a hive up to maximum strength, you've got to supply them with a carbohydrate stimulus. It takes pollen and carbohydrate—sugar syrup, or corn syrup, honey, something. You can buy a sugar source, but you can't buy pollen. That's what you're really going south for, you're going south for pollen.

"That queen, in Orlando or in Boston, she's going to start laying at the same time of year, either place. In New England they start brood rearing in January and February, and then when March rolls around, they don't have the pollen. You have your biggest losses then. Colonies die, from a lack of protein. For a commercial beekeeper it gets scary because you don't know what kind of numbers you're going to have."

There was a bee smoker on a pallet nearby, and Andy lit it. He wanted a look inside a hive. He blew smoke in the entrance, through the hole in the cover, and without veil, gloves, or even a hive tool (used to pry up frames), he lifted out a frame. It was covered with bees. He blew a stream of air at it. The bees on the comb made a shivering sound and cleared away from this blow of breath, leaving an open expanse of honey-brown waxen quilt, which covered pupating brood.

"That's a good pattern," he said. "That's good brood." He looked at another comb, and this too was filled with nascent bees. Andy's enthusiasm showed. The prospects looked good.

But when he walked to the feeding tank, where Jeff and Dale were still pouring syrup, and Jeff told him about the leak in the elbow joint, Andy became nervous. The draining syrup was like a drain on the business. Talking aloud, he said he had to get all these hives inspected so that he could get twelve hundred hives into New Jersey by April 15—bees circling, landing on our shirts—and there was no telephone in the trailer, and he couldn't call Crystal, and a hardware store was twenty miles away, and there was more syrup to get tonight, and the bees had to be fed antibiotics, and Dale better go to the hardware store and he, Andy, had to get going because it was a five-hour drive to the inspector's office at Clemson University. Andy kept talking even after he got into the car and closed the door.

A season was riding on the pollen and sugar interlock. And the state of the hive economy had a connection to the state of Andy's inner and outer economy, his spiritual and material life. He couldn't stand to see this carbohydrate energy running into the ground. It was not a good sign. Beekeeping, primitive art that it was, despite the introduction of twentieth-century technology, ultimately came down to a matter of good and bad signs.

"'Bye," Jeff and Dale said in unison, and Andy drove away. The dust from the road settled. Sounds emerged, glass jars hitting together, the sporadic voices of the two workers, and the continuous voice of thousands of tiny wings moving air.

9

After World War Two ended, Andy Card, Senior, went to Boston University on the GI bill, got married, and worked as a salesman. He looked for something to do in agriculture, too. When he was a boy in Maine, Card's neighbor had kept bees, and he had watched them. Now Card remembered them, and he began like most other beekeepers, with a hive in his backyard.

In 1951, the year Andy, Junior, was born, Card put four hives on a car trailer and moved them west of Billerica to the Nashoba Valley, an apple-growing region in central Massachusetts, and rented them to a grower. Some growers had noticed that a good supply of honeybees made for fewer lopsided apples. Card worked this into his pollination sell.

In the 1952 season Card had forty hives. He wanted four hundred. He bought defunct operations—hives from a widow, hives from a retiring beekeeper, and hives from growers who couldn't manage bees. He spent his nights putting frames together and assembling hive parts.

Although Andy Card, Senior, always had a job outside beekeeping, his son took no other work. Andy, Junior, became a beekeeper in 1953, when his parents entered his name in a raffle at a beekeepers' club meeting. Bees were on Andy's mind at an early age, he says; he was told that his first word was "bees," spoken when he had a cold and his mother applied a menthol salve to his chest, because the sensation reminded him of a bee sting.

By the time Andy was eight, his father had set up an apiary with twenty-five colonies and told Andy the yard was his to maintain. At twelve he read *Honey Getting*, a book that advocated reversing the brood chambers. The idea is that the colony tends to move to the top of the hive, where it's warmer. Often in a two-story hive there is a near-empty box on the bottom, so the beekeeper switches the two. The queen moves up into the empty box and builds a nest, the bottom box hatches out, and the population soars. The peril is that the reversed nest, close to the entrance, will freeze and die, or "chill," so the beekeeper, like the gardener, must not act too soon.

That year, 1963, Andy got paid for the first time, $10 for forty hours' work wiring frames. By 1965 Merrimack Valley Apiaries had two thousand hives and was the biggest beekeeping company in New England. Andy was hiring his high school friends to help hand-load the hives on trucks for the moves to Maine.

In 1969 Andy's mother died. His father lost interest in the business and didn't bother to hire a foreman. Andy, in his grief, decided to take over. As beekeepers can do, he made life where there was no life—everything that was a box, he put bees in it. He had, he says, "more increases than wits," and got the MVA colony count up to twenty-six hundred hives. His high school friends worked full-time, driving the trucks, delivering the bees—eight hundred hives into apple orchards, sixteen hundred to Cape Cod, two thousand to Maine. Everyone worked for minimum wage, and MVA had its biggest profit margin ever.

The same year, DDT having been banned, farmers began to spray

Sevin on corn. Corn produces a large amount of pollen. Just after Card put six hundred hives in a field in Maine, hoping to get a honey crop on alfalfa, a neighboring farmer sprayed Sevin on corn early one morning (rather than the safer time of dusk), just as the bees were taking to the air. When Card opened the hives, dead bees rolled off the combs.

A senior in high school, managing a large business for the first time, Card saw his stock dying. More bees died when he took the colonies home, and they continued to die through the winter. MVA went into the winter with twenty-two hundred hives and came into spring with five hundred. For several years Card split and fed to increase colony count, and bought packages of starter colonies, but the field bees were weak, and the growers knew it.

Then in 1973 Tom Charnock drove into the Maine blueberry barrens with hives just out of orange groves and hot as pistols. And not only were the colonies strong, they were on pallets. The blueberry workers in Maine didn't have to spend late-night and early-morning hours lugging Charnock's hives from a truck bed into a field; they could use Charnock's forklift. The blueberry foreman, fifty-five, young and vigorous when MVA first arrived in the barrens, seventy and tired when Florida bees arrived, took to driving by the MVA loads and going home to bed.

Between 1974 and 1976 MVA nearly lost its accounts in Maine, accounts that amounted to 40 percent of its pollination rental. The Cards were still suffering from pesticide kills, and their honey crops were down. Andy wanted to go to Florida, but his father, a traditionalist, was skeptical. It would be expensive. The debates between father and son continued for several years, but migration was inevitable.

It was in 1977 that Andy Card, Junior, drove his first load of bees from Massachusetts to Florida. Traveling through Georgia on Route 301, a two-lane highway, Andy suddenly came upon a police cruiser that had stopped a car and parked in the traffic lane. Carrying many tons of wood, wax, honey, and bees, he had no chance to stop. A van was coming the other way. Card drove heroically, especially for a twenty-six-year-old beekeeper making his first long-distance trip in a semi. He navigated the opening, but he nevertheless radically rearranged the positions of the automobiles.

No one was hurt, and though he was clearly not in the wrong, Card was questioned through the rest of the night. What was he carrying? Where was he going? What was he going to do there?

When daylight came, when the bees were about to leave their hives, Card was allowed to move along.

10

Aristotle noticed that a bee, finding a nectar source, would return minutes later with its hivemates. When a bee finds a profitable nectar source in the field, it may advertise its find. The bee returns to the hive and gives samples of the nectar, often in a defined territory of the hive. Having attracted attention, the bee also performs a "vibratory episode."

If the nectar source is within one hundred yards of the hive, this episode consists of a spin or run in a tight circle. The bee shakes its abdomen quick as a blur, and turns clockwise. One circle completed, the bee stops, collects itself, turns and thrums counterclockwise. Its enthusiasm during this vibratory episode often correlates with the profitability of the nectar source, according to the work of Karl von Frisch, who studied the honeybee for many years and received a Nobel Prize for his deciphering of the inner hive.

Watching this round dance, sharing in the discovery, other bees become equally enthusiastic. The nectar sample helps the bee in the field, or if there's no sample, the follower bees feel the forager with their antennae, picking up scent and taste from her body. That can be enough, and with the scent in mind, they eagerly seek its source.

Of course the round dance is not necessary when, after a flash of light and a slam like thunder and an unrecognizable gurgle, nectar sprinkles down into the nest and over the backs of the bees.

11

Dale Thompson and Jeff Kalmes had moved the syrup truck into the woods, eluding the bees for a while, but now Jeff was squinting to keep them out of his eyes and Dale had to keep brushing them off his clothes.

Jeff was twenty-four, with apple cheeks and blue eyes. He was a tireless worker, and had stories of strings of twenty-four-hour days loading and delivering bees. At the back of the truck, he stopped for a moment to mix some antibiotic powder into a barrel of water with a two-by-four. The tool didn't suit him. "This is the trouble with being in too many places," Jeff said. "You can only carry so much stuff. The main operation is up north, and the rest is in Florida. But none of it is any good to us now." He nodded at a stack of seventeen hundred honey supers stacked on a trailer bed. "Someone could come up here and drive that trailer off. Or burn it. Beeswax burns real good. You ever seen a hive catch fire? I saw a bee yard burn once. I was working for a woman who keeps bees in Florida. She set a hot smoker down on the ground and a grass fire started. She had treated her hives with paraffin, to protect the wood, and once they caught, there was a ten-foot flame. We pulled the straps off the pallets, and carried some of the hives away. It took a long time for the fire department to come. The shirt burned off my back. The next day there were a hundred twenty little round plies of ashes, the pallets we didn't get."

I knew Jeff as an athletic forklift driver. He could raise a pallet on the run, like a shortstop scooping a grounder, loading trucks faster than anyone I'd seen. Jeff delivered the loads to Nantucket and Blue Hill, Maine. In the dark, maneuvering over rocks, he would set out three hundred hives, putting them far enough apart to cover the area. If he hit a rut and a pallet fell off, which happened regularly, he had to stop and put the hives and the pallet back together in the dark. Blue Hill was a popular habitat for bears, and sometimes the bears would tear open half the hives. On the return trip Jeff would have to clean up this bear damage.

After he finished high school, Jeff applied to a beekeeper's college in Ohio, but then decided to work for commercial operations in Georgia and Florida. Jeff liked to tell war stories. He had taken stings in the ears ("rings for hours") and under his fingernails ("the worst place of all, stings for days"), and one time, when he was on top of a truckload of hives, a bee climbed up Jeff's pants and stung him where no man should ever get stung. He fell to the ground.

Dale dipped the jars into the antibiotic mix before Jeff filled them. The antibiotic was to prevent nosema, a bacterial disease that's something like diarrhea. Nosema can debilitate a colony, especially in winter, when the bees are confined. Colonies with infections of nosema are easy to identify because of the stains left at the entrances. Nosema is one reason some beekeepers wear hats. Dale, wearing a visor cap, told a hat story. He had been in a pasture in Florida working some beehives and had run out of fuel for his smoker. There were no pine needles, his preferred smoker fuel—no fuel at all, in fact. So Dale was left with two choices: burn his socks or burn his hat. He burned the socks.

In the preforklift era, Dale said, "we'd carry hives all night long. I'd wear a bee suit, but I wouldn't wear gloves. It was just too hot for gloves. By the time morning came, my hands looked like potatoes. You'd get up to Maine, and the farmers there knew nothing about bees. They thought you got the best pollination if you put the hive right next to the tree. You couldn't tell a farmer the bees will fly two miles. You'd have to follow him into the orchard, with him telling you, 'Point it at that tree. I got good apples from that one last year.' "

Dale was big and loose-jointed, with a ready, broad smile. "I learned how to drive a semi backing around the bee yard. I'd drive fifteen, forty-five minutes at a time. And then I'd practice my parking at the bee yard, I'd set up supers as parking spaces.

"One year I put forty thousand miles on the truck. That was a part-time job, since I had a full-time job as a beekeeper, too. It gets to you. A couple of times, pulling out of the truck stop I had to look in the rearview mirror to see which way I was going. If I had bees on the truck and it was springtime, then I knew I was going north. I'd lose track of time. You can't stop because it's hot and the bees will cook. So you get out and run around the truck a few times, or pour ice water down your back. It's like a game: see if you can finish. I don't know if that's the way life is supposed to be, but I just like to finish."

Jeff and Dale were thinking about finishing beekeeping. The mite crisis and the depopulations had made them question their place in the industry. Despite the pleasant aspects ("a hundred sunrises a year"), the excitement of the travel and adventure of migratory beekeeping had worn off. The boys, as Andy affectionately called them, were getting older. They both were thinking about getting married. Jeff's girlfriend didn't understand. As Dale put it, "Girls are on schedule with the rest of the world. Andy has Crystal, but girls I know are not keeping bees."

A pallet of hives with the ideal population for pollination. —GLENN CARD

Andy arrived with Wesley. He gave Jeff and Dale directions to some other yards and told them how much feed they'd need. Then he called to Wesley, who, wearing a beekeeper's hat and veil, was playing with a toy truck in the dirt, and he called to me. We left Millbury City.

We drove along Route 301 to the sites where Andy had placed his hives. Andy was an optimist by nature, but his talk was of the difficulties of the year. He wondered if there would be enough pollen here in South Carolina, if the bees would get strong enough to pollinate well and satisfy growers, if he would stay mite-free, if he would even be able to get into New Jersey, where there was a bee inspector who was trying to keep migratory beekeepers out of the state.

We turned off the road, drove by an abandoned house along the edge of a hayfield to a hollow where Andy had placed about a hundred hives. Andy got out of the car and crouched down for a look at the flowers, picked one, smelled it, opened the petals. "Not a bad location," Andy said. "They may do well here." He walked by the pallets and looked at the hive entrances. "One-third good, two-thirds marginal." He stopped and took an admiring look at a hive with a cluster of bees bulging from the entrance; this was the "beard" that Andy liked to see.

"We'll split a thousand hives, starting Sunday."

We drove to another hayfield where there were another hundred hives. A strip had been plowed close to the hives, and Andy guessed that the farmer who owned the field was about to burn and had built a fire-break to protect the bees. Burning was common here, and trails of smoke were visible off in the distance.

More beards on these hives. "Surefire splitting the most advanced hives here," Andy said. We walked back to the car, where Wesley was waiting for us. His head came up to the top of the front fender. His face was sunburned. He said he was thirsty. One more stop, Andy told him.

We drove about ten miles and then pulled over by a field that was being plowed. "Damn," Andy said. "Watermelons. They didn't tell me they were gonna plant watermelons when I got this location." In December, when he had put the hives at the back edge of the field, a hundred yards from the road, it had been clear sailing. Now there were furrows in the dirt three feet from crest to trough, and there was no apparent way to drive a truck across the field. One good thing—this farmer would have plenty of bees to pollinate the watermelon flowers.

It was a slow walk in, through the freshly plowed but dry and soft sand. Wesley had a hard time of it, and before we were halfway in, Andy told him to stay behind. Wesley held on to a tree branch.

"How am I going to get in here to feed?" Andy asked. Not that I could tell him; he was just thinking out loud. He looked through the brush for a trail. "And the other question is, How am I going to get these out of here? We've got to go to Jersey by April fifteen."

Since the hives hadn't been fed, and rains hadn't brought on a nectar flow, there was no carbohydrate stimulus, and the entrances to the hives were barren. Once again, with no smoke, no veil, Andy walked up to a hive and without hesitation pulled off a cover. He looked in. He slapped the hive shut, opened another. This induced bees to take to the air. They found me. One bee tangled in my hair, then another. I tried to hide behind some tree branches, and spun around in my own kind of round dance.

Andy was grabbing at his hair, too, but just to pinch the offending bees. "We've got to feed these hives," he said. "I got to talk to that farmer, find out about this." Opening, closing hives, bees dancing in the air. "Jeff has to get in here somehow."

Wesley was waiting for us along the edge of the field. His cheeks were bright red. Andy lifted him to his shoulder.

"You're tired, aren't you Wes?"

"Yeah, I'm tired. I want some milk."

"We'll get you some."

"We got to pull our belts in," Andy said, turning back and looking at the stranded apiary. "It's going to be a rough go ahead. We've got to cut down on expenses somehow."

South Carolina, because of a cold snap and a lack of rain, didn't produce the way Andy had wanted. The pollen flow stopped short and his colonies weren't building. Filling the syrup tank again and again, Andy continued to feed, but the protein wasn't there, and he had to boost the population in his hives.

Instead of tightening his belt, Card loosened it. He invested $20,000 in package bees. He bought them from Reggie Wilbanks, the package producer in Claxton, Georgia. The sum bought a thousand two-pound packages—a ton of bees—and a ton of bees, Andy hoped, would strengthen the weaker hives for the move north.

12

In Claxton, Georgia, Wilbanks Apiaries was readying for their busy time, when they worked seven-day weeks and twelve-hour days making packages—two pounds of bees and a caged queen—for beekeepers who would use them to make hives. The package season would come in April if the mite crisis didn't spread into Georgia and cause an embargo. When Reggie Wilbanks took me to his yards, he was gearing up for a full season.

Wilbanks Apiaries had been selling fifteen thousand packages of bees a year—thirty thousand pounds, fifteen tons of honeybees. And they sold fifty thousand queen bees to beekeepers in the United States, South America, Europe, the Middle East, and Asia.

Tall, with a soft, honeyed voice and a talent for politeness, Reggie Wilbanks was a beekeeper's beekeeper. He grew up in his father's and grandfather's company, had fifty hives of his own in junior high school. The beehives paid expenses at Georgia Southern, where he received a bachelor's degree in industrial management in 1972. Reggie inspired the kind of admiration that made you want to elect him to something. He had been president of the American Bee Breeders Association, a director of the American Beekeeping Federation, president of the Georgia Beekeepers Association, chairman of the Georgia Farm Bureau Honeybee Commodity Committee, president of the Claxton Jaycees, vice president of the Evans County Farm Bureau, a member of the National Mite Advisory Committee, and a vice president of Wilbanks Apiaries. And Reggie worked in the field, handling bees with the other workers. He believed it improved company morale. He even, as a matter of tradition, poked all the nail holes in the fifteen thousand lids of the sugar syrup cans that were the food source for the bees in the package. "That way if there's any failures—holes too large or too small—we know who's responsible," he said.

At 7 A.M. we left the company warehouse—seven men in two trucks—and drove for twenty minutes. We came to a clearing in the woods where the oak leaves had been raked to gray sand, and there were fifty hives set in three rows, on concrete blocks. Here, and at the dozen or so yards that followed, Reggie's crew collected about five hundred stock boxes. A stock box was a screened cage the size of a cigar box into which the beekeepers shook twelve to fourteen ounces of bees, about two thousand bees. Each box would become the support colony for a virgin queen.

The crew filled the boxes by holding frames over a funnel and giving them a hard shake, with the bees rolling down into the stock box. They measured by eye; half full was about right. They collected one box from each hive, and they clamped the boxes end to end into racks a foot wide and eight feet long. The finished racks were then clamped into racks on a flatbed truck. The startled bees fanned their wings; the cumulative sound was like a wind moving through grass.

"You have to tip your hat to the people who work for us," Reggie said. We were at the edge of the clearing, watching the crew work. "If I didn't grow up with bees I couldn't imagine wanting to go out and get my ass stung off every day. It's a M.A.S.H.-type situation, you know what I mean?

If they get stung on the eyelid, you make fun of them. They do the same to me. As long as you have the right attitude you can have a good time."

From time to time Reggie left me, filled a watering can with sugar syrup, and passed by the workers, pouring syrup into plastic troughs inside the open hives.

"I work a lot of young guys," he said. "They usually come out either because they want to work outdoors, or because they need the work. Usually it's because they need the work. The new ones work alongside the ones who have already lost their fear. They see them and it becomes a matter of pride. 'He can do it, I can do it,' they're thinking. Losing fear is one of the reasons people become attached to bees."

The bees the men were shaking were nurse bees and house bees, not field bees, so there was not a great deal of stinging. House bees cluster by nature. Shaken from the combs, some took to the air and landed on and crawled over the beekeepers. One man had a curtain of bees hanging from the back of his hat brim, a foot-long cluster that shook and shimmered when he moved. Chewing tobacco, digging into hive after hive, he ignored the braid of bees hanging down his back. At the end of the yard's work, another beekeeper came by and shook the man by his shoulders—the man staggered, his arms went out, and the cluster, failing to the ground, broke apart and lifted into the air.

"You've got to be careful when you're shaking bees," Reggie said. "It's like a sack of eggs. The membranes in the honey stomach can break if you're rough. Some other producers make bulk pourings of bees. They wet them down and use a dipper to put them into cages. That's brutal, mean. You come back, they've all absconded. They're all up in the trees."

Absconding, or abandoning the hive, has always been a problem for beekeepers. A migratory beekeeper from Florida once paid a beekeeper in North Dakota $10,000 to shake bees into several hundred hives. The bees were shook and the hives were loaded on a truck and the driver left for Florida. But along the way, because of the heat, the bees left the hives and pressed against the nets, which puffed out from the truck. In the beekeepers yard in Florida, when the nets were pulled off the truck, most of the bees absconded. As the beekeeper put it, "Ten thousand dollars flew into the trees."

There was one beekeeper in Reggie's crew who was known as Beeman. Charles Kennedy was his real name. Beeman was twenty-nine, and he'd been working for Wilbanks for fourteen years. None of Reggie's crew worked with gloves—Reggie figured they would be more gentle with the bees without gloves—and Beeman didn't wear gloves either, but he liked the roiling air, liked roughing up a hive and taking a few stings. His arms were scarred from stings. He'd take a hundred stings, some days. "Charles stirs them up a bit more than he needs to," Reggie said.

Before the crew shook any bees through the funnel, they had to first find the queen bee and set her frame aside. Without the queen the colony would falter and stop reproducing, until she was replaced by either the bees or the beekeeper. To find the queen, the beekeepers lifted frames and ran their eyes over the thousands of bees. Sometimes they emptied the entire box before they found her. Though she was markedly different from the worker bees, looked more like a long brown wasp, she was difficult to spot among the thousands of other bees.

Beeman prided himself on his quickness at finding queens. Sometimes they held queen races—five hives, five queens, for a few dollars or for beer. One time, the story goes, there was a match against a beekeeper from Mississippi. "He said he could find some queens," Beeman said in a soft, snarly voice. "So we went looking for some queens. I bait these guys, you know, lead them on. I talk to them, say, 'You're going to win,' like that. I keep talking so the guy will be listening, and not finding queens." Beeman won a hundred dollars.

Beeman walked by, carrying a queen between his thumb and forefinger, holding her by the wings. "Paint job," he said. He opened a tool box with his free hand, picked out a bottle of fingernail polish, and put a dab on her thorax. She wouldn't be as hard to find next time. He dropped her inside the hive, closed it up, and moved on.

13

When beekeepers in this country used the black bee of northern Europe, they increased their hive count by swarm getting. Beekeepers kept bees in small chambers to force them to swarm, then caught the swarm. In some areas laws permit the beekeeper to trespass while in pursuit of a swarm. Colonies were killed off at the end of summer and the honey was pressed from the combs.

Large-scale beekeeping was not possible with the black bee and the swarm-getting system. Black bees swarmed too readily. One old beekeeper who had worked the black bee in Massachusetts (it survives, they say, in isolated pockets) told me that a colony would swarm if you lifted the hive cover, or if you sneezed in the bee yard. Their population diminished, swarmed colonies don't produce surplus honey crops.

The Italian bee adapted to the long flowering season and the mild winters of the Mediterranean region. It did not swarm as frequently as the black bee. It tended to stay put, make wax, store honey, and make large populations—fifty thousand, eighty thousand workers—and honey crops weighing one hundred or even two hundred pounds.

During the 1860s one Italian beekeeper capitalized on his stock and shipped queens from Venice for $20 each. Some American beekeepers traveled to Italy to purchase queens, and for a time Italian queens were prized and very expensive. One sold for $22.50 in Iowa in 1860. But technology for queen propagation developed quickly, and supply soon caught up with demand. In 1869 Henry Alley, a Massachusetts breeder, sold Italian queens for $2.50 each. That price stood for a hundred years.

Gradually the black bee gave way to the Italian golden. But the Italian race had not adapted to long winters. The large populations tended to starve, and beekeepers who had suffered winter losses needed new stock. In 1879 the editor of a beekeeping magazine offered a reward for a quart of live bees shipped to his address. A few beekeepers tried but none succeeded in keeping the bees alive during transit. In the 1880s cages and a means of feeding (a can with holes punched in the top) were developed, but the package industry didn't take off until sweet clover, with its rich nectar flow, spread into the farming regions and prairies of the Midwest and West just after the turn of the century.

The package industry soon shifted to the warm southern states, with their advanced spring and early pollen. In 1913 a company shipped several thousand packages from Georgia. When the railway services improved, the industry expanded.

The package industry grew again when beekeepers in Manitoba, working the plains, showed that a package of bees installed in the spring could make a two-hundred-pound honey crop in the same year. A new type of annual management developed, a throwback, really, to the days of swarm getting. Canadian and plains states beekeepers killed off most or all of their bees in the fall, took the honey, and in the spring installed new package bees from the South or California.

14

Reggie was working two crews—one gathering stock in the field, and a second one making queen bees at the queen yard behind his house in Claxton. It was a tightly packed, highly organized apiary, with three groups of hives around a workshop.

Inside each breeder hive was a prize queen, and each day a comb of her eggs was taken and placed in a starter hive. The starters were the second group, two rows of hives with a great number of young nurse bees. The eggs hatched in the starters and remained there two days, during which time they were fed copious amounts of brood food, a complex of fluids secreted from the honeybee salival glands. Nurse bees dribble brood food down the cell walls until the larvae float in it.

The combs of two-day-old larvae—each only a few millimeters long and at the prime age for maximum ovarian development—were removed from starter hives and taken to the workshop for grafting. The grafting house was a kind of way station in which the larvae changed beds. Here they were moved from worker bee cells and laid into queen cups, waxen cups a quarter-inch across. In a queenless colony, workers would add wax to the cups, "drawing" them out into finished queen cells—and the larvae would develop into queen bees

In the grafting house a broad-backed, red-haired man was bent over a comb of larvae, lifting them out with a tiny spoon. His name was Dennis Oliver.

"How many queens would you say we put out here, Dennis?" Reggie asked. "Fifty thousand?"

Oliver was working a toothpick between his teeth. His eyes were red.

"I'd say it's more than that, Reggie."

"Dennis has a ninety percent take on his grafts," Reggie said, meaning that nine out of every ten larvae that Oliver grafted into queen cups developed into mature queen bees. It seemed remarkable that it could be done at all. It certainly demonstrated the resilience in the bee colony.

Oliver worked quickly, like a cook spooning soup, lifting larvae from comb and setting them into queen cups. There were twenty cups affixed to a wooden bar. Three bars were mounted into a standard hive frame. Once loaded, the frames, each with their sixty cups, were taken outside to the third and largest of the hives in the queen yard, the cell-builder hives. Each of these sixty hives was highly populated, highly fed, with a queen confined to the lower box. This made for a queenless disposition in the upper box, and the nurse bees kept building queen cells. There was a nine-day rotation scheme among the cell-builders (the Doolittle system, it was called, after the man who developed it). When Dennis Oliver and his helper put a newly grafted frame in a cell-builder hive, they took a frame of finished queen cells out.

It takes sixteen days for a queen to develop, and when she is removed from the cell-builder, she is fourteen days along (two days hatching, three days in the starters, nine days in the cell-builders). That leaves two days for the beekeepers to place the cell in a hive. On the sixteenth day she emerges, looks for other queen cells, and destroys her rivals.

"Almost anybody can make queens," Reggie said. "The difference is the little things you do to make a quality queen. Thickness of feed, type of swarm box, amount of bees in the box, and timing. The real key is timing."

It was a straightforward process, really, a matter of organization. The mystery, of course, is in the hive, in the process of trophogenesis. Queen bees are created by specialized feeding—compounds that determine body type and caste, and whether a larva becomes a worker or a queen. All larvae are fed great amounts of brood food and royal jelly, but worker larvae receive the royal jelly for only two or three days. A queen larva receives royal jelly for her entire six-day larval period, and this accounts for her different body. On the last day of feeding alone, before sealing the queen cell, nurses may make thirty-five hundred feeding trips to a queen larva.

Next to the bench where Oliver worked was a food freezer that had been converted into a holding tank for queen cells. It was heated by a light bulb. When Reggie opened the freezer, he found two queens that had hatched out and were walking along the walls. They would, perhaps, soon seek rivals.

Reggie picked one out and held her to the light. "That's a fair queen," he said, turning her over, examining her. "Good size to it." But then he crumpled her, and tossed her to the floor. He picked up the other queen and said offhandedly, "This one's a little better," and as an afterthought, "They darken with age."

I stiffened, wanting to ask him not to do it. I imagined a hive generated from her. She was worth $6. But I didn't speak, and Reggie crumpled and tossed. It had to be done; they would have eventually torn up the other queens. These queens, two of many thousands, meant little to Wilbanks. His business, after all, was predicated on the practice of requeening, or killing off the year-old or two-year-old queen and replacing her with a more vigorous one. He was doing the right thing, but I'd been a different kind of beekeeper, an amateur, by definition an idealist, one of those who was moved by the sight of the queen bee running the comb. She had that opulence, that ovulance, that mystery that suggested not only a link with human life but, with her long leather-colored body, other links as well, perhaps with the aboriginal wasp the honeybee had emanated from. She suggested the angiosperm, explosion and all the eons before it.

Each day at this time of year four hundred to six hundred stock boxes were gathered and four hundred to six hundred queen cells were pulled from the cell-builders. The two groups came together at the end of the day in the queen mating yards. Here the queens hatched out of their cells, went on their mating flights, and a few days later began to lay eggs.

The queen mating yard was in a clearing, under longleaf pines. The late afternoon air was cool and pleasant, with the long pine-needle clusters turning black with the setting sun. The hives in the mating yard were small "mating nucs" (for nuclei), not much wider than a shoebox and slightly more than a foot high. They were arranged in long, curving rows, a broken pattern that helped the queens memorize their locations when they left to mate. There were two hundred and fifty mating nucs, but since each box had two chambers and two entrances, was in fact two colonies in one hive, the yard contained five hundred mating colonies.

Each hive had three frames, and a small entrance hole the size of a quarter covered with an aluminum disc. The crew first gently pressed a queen cell (carried in an egg carton) between two frames. Then a stock box was opened and the bees were shaken over the frames. The stock box, the same size as the top of the chamber, was left in place until later in the night, when the crew would open the entrances. At that time the bees would, Reggie said, "run out on the front porch and look around." Spending the night together, mingling scents, they would by morning be a settled colony and probably not abscond.

There was a big hive at the corner of the apiary, and inside it was a

A hive undergoing inspection for brood production before pollination.

—GLENN CARD

large proportion of drone comb, which would produce the mating stock. There was a lot of flight coming from this hive. Drones flew in specific areas of the sky, about thirty-five feet up. Drone zones, these were called.

After hatching, and feeding, and a period of restlessness, the queen takes to the air for her mating flight during the hottest part of the day. She seeks drones, instinctively knowing where they are, but sometimes she has to fly several miles to find them. The distances are immense, when you think of the proportions—a bee, less than an inch long, flying several miles of unknown territory.

When the queen finds a drone zone, she courses through, trailing pheromones that are irresistible. Hundreds of drones pursue her, a comet of bees. The fastest drone has access to her, and they couple at top speeds. The queen then veers away with a loud sound that precedes the drone's last slow drift to the ground. There is a succession of matings, up to fifteen, after which the queen has collected enough sperm to last her laying life.

When the queen returns to the hive, she does not fly again, unless she leaves with a swarm. It's incredible that she makes it back at all, that she is able to memorize the position of her hive among all the others—to memorize the landmarks, the trees that mark the way, the position of the sun in relation to her flight line—and to reverse it all on the return trip.

There were five hundred colonies in this yard, so five hundred queens would take to the air for mating flights almost simultaneously. And Reggie had ten thousand mating nucs. The air in Claxton must have been a spectacle, a kind of honeybee aerial fair, with so many virgin queens and drone comets coursing through.

A few days after her mating flight the honeybee queen begins to lay a thousand or perhaps fifteen hundred eggs a day—enough eggs to equal her body weight. Can you imagine a chicken laying a dozen eggs a day? It would be hell in the barnyard.

Two weeks after joining the stock boxes and the queen cells, the Wilbanks crews would return to the mating yards. Reggie, Beeman, and the others would check the combs and look at the laying patterns. If the queen was productive, they'd put her, along with a few attendants, in a cage the size of a matchbox—and that matchbox would go to Canada, or Wisconsin, or Brazil, or Trinidad as a single unit or in a package.

And after taking the queen out, they'd press another queen cell between the frames, and the process of hatching and mating would begin again. It was all, really, an alteration on the ancient and natural scheme of reproduction and colony fission.

15

Horace Bell's bee ranch was cut out of some woods of live oak in Deland, twenty miles inland from Daytona Beach. The road in passed an orange grove that had last blossomed in 1983. The freeze of 1985 had finished those trees off, and the leaves hung like thin leather. Only the dusty live oaks remained, and the Spanish moss, the wrist-thick vines.

Horace built his house with cypress trees he cut himself. His office was on a porch, where he could see the workplace and warehouses. At the other end of the porch was a swimming pool, covered with a screen cage to keep the bugs out. Horace began his workday at five, and usually he had made a number of calls before seven, when the crew arrived, punching in at the time clock on the porch. Most people knew that the best time to call Horace was at 6 A.M. He was in bed most nights by eight-thirty.

Horace kept to himself. He worked seven days a week, usually, and other than his family, Horace's only interest was bees. He thought about them, he said, twenty-four hours a day. Horace had that trait of preoccupation that drives a person to get a beehive. Until 1985, when Horace realized he had to make contacts in the beekeeping industry and began to go to meetings, most commercial beekeepers didn't know what he looked like or how old he was. Most thought he was an old man. They were surprised that he was still in his thirties.

Horace was talking with three bee inspectors when I arrived. He thought it amusing that he had to give the inspectors jars to put the bees in and tell them where to find the hives. Finding Horace's bees wasn't a simple matter, since they were spread out from the Florida panhandle to Miami. "Nina will show you around," Horace said to me, and to Nina, with a slight smile, he said, "Don't give him all the good stuff." Nina Kelso was Horace's oldest daughter, eighteen years old and in the business. Nina and her two sisters had worked in beekeeping since preschool days, and got so good at assembling frames that Horace set the piecework wages according to the pace of his girls.

Though the temperature had been up into the nineties for the last week, it had suddenly dropped into the forties. In some ways that was good—it would slow the bees down. Nina had put on two flannel shirts. She was a tall, pretty woman, with her father's high cheekbones and her mother's spray of freckles across the cheeks.

Nina led me into a woodshop and pointed to a pile of newly assembled frames ten feet high. In the next room was a five-thousand-gallon tank of corn syrup. Near the tank Horace's employees were sluggishly pulling on bee suits.

I asked Nina if this was her true career choice. "I'll probably be doing this for the rest of my life," she said. "I went to school for a while, to study accounting, but I came back here to work the bees. I do some of everything. I drive the forklift, the trucks. I've gone up to North Dakota the last two years. I'd come back when school started. I'd unload the trailer trucks at night when the bees came back. Sometimes I'd just finish, take a break, and then I'd get a call from Dad. There was another load of bees coming in." (Horace had sent nearly twelve thousand hives to North Dakota.) "The worst part is when they fall off the forklift."

"What do you do then?"

"Get off and pick them up by hand."

We walked through the extracting room. Horace had two large extracting units, each with a mechanical knife, a conveyor, a spinner basket, pipes, tubes, and filters leading to settling and storage tanks. At the end of the room was a screen and light where bees collected during the extraction process. Most beekeepers shook or blew bees off the combs in the field. Horace didn't. He reasoned that the field bees would, after the honeyflow, sit around the hive and eat honey. So he brought them home. Hundreds of pounds of bees collected on the screen. The crew scooped them into gallon jars and poured them into newly made hives. It was a way of getting more out of the bee.

Nina took me into two warehouses, each with 240 hives inside. It was dark, and the bees weren't flying. They don't, usually, inside a warehouse. The hives would be split that day, 480 hives becoming 960.

She showed me two other warehouses. One housed thousands of hive bodies, and in the other were columns of steel barrels that reached to the rafters. Horace sold honey not by the jar but by the fifty-five-gallon drum. A drum, or barrel, holds 660 pounds of honey. When Horace had started out in beekeeping as a teenager, he got $60 a barrel. These days he was getting about $420.

I asked Nina if she was going to North Dakota that summer. She turned and smiled, and I saw she was pregnant. Her outer shirt had concealed the obvious. "Six months," she said. "I'm due at the end of May." The bees would go north, she said, but the family would stay.

"Do you want to see the queen incubator?" Nina asked. We went to the porch, to a tall gray cabinet. Heat drifted out when Nina opened the door. Inside the cabinet were egg cartons filled with queen cells, some from Wilbanks Apiaries. "Two hundred are supposed to hatch today," Nina said. "They have to go out to the hives today. Five hundred more just came in. If we don't get them out, they're going to hatch in here and start running around."

There was a palpable stillness there, a sense of emergence and of work to be done. Seven hundred queen bees could produce seventy thousand workers in a day, a million in two weeks—bees to forage in the clovers of North Dakota. If the quarantine ended.

16

"This is just not a good year for orange-blossom honey," Luella Bell said. We were on Route 44, central Florida, passing through a defoliated landscape. There were long views of rolling land, lined with dead orange trees. Some trees were standing and covered with moss. Others were cut off at the stump and painted white, like strange lawn decorations. Some trees had put out new branches, but those had burned in the most recent freeze. Pastel houses that had been deep in the groves were now exposed and looked scandalous. It was a sad sight. Orange trees are among the most resplendent of trees, with glossy deep green leaves, with bloom and fruit on the branches at the same time, and with a fragrance that can perfume an entire landscape.

The worst freeze of the century in Florida came on Christmas Day, 1983, when the temperature dropped into the upper teens. Cold winds from the northwest held the temperatures down for several days, and about 160,000 acres of citrus trees were lost, 10 percent of the trees in

Florida. Some of the trees recovered in 1984, but in January 1985 another series of freezes defoliated those and many other groves.

"Orange used to be our main crop," Luella said. "That's gone. And we used to go to gallberry, in north Florida and Georgia, but a lot of that froze, too. Horace and I don't know how we'll make it without a honey crop until next winter, how we'll carry it all. We'll have to start laying off.

"There is orange farther south, but a lot of the growers won't let you in because of the orange canker." Orange canker was a tree disease that had been eradicated in Florida in 1927 but then reappeared in 1985. Some growers were reluctant to let beekeepers in the orchards for fear they'd spread the disease. "In some places they spray the underside of the truck with herbicide and let us in. We try to be as helpful as we can so they'll let us keep coming."

Clover and sunflower in the Dakotas followed orange and gallberry. "There's been snow in South Dakota," Luella said, "which is good for clover. They figure it's going to be a pretty good year up there, if we can get up there."

Word of the extent of tracheal mite infestation in Florida came when the Bells were in South Dakota. Georgia had passed an embargo on transport of bees. The return runs became covert operations. One driver, Luella said, was stopped in Georgia and had to pay $800 to get out of the state.

"We were constantly moving bees last summer," Luella said. "Seems like you no sooner put them down in one spot than you picked them up and moved them to another spot. There was a poor crop on clover in South Dakota. We had somebody looking for us, someone running around getting signatures so we could put them down. When he said the sunflowers in North Dakota were opening, we just moved as fast as we could. Once they're open, it's two or three weeks and that's it, they're gone. You got to get right on them."

In South Dakota, Luella drove the forklift. She needed to keep moving, she said. Horace did the extracting. "The thing that really bothered me up there was living on a couch, not having a bed. All that time and no place to put your feet up. It's that sort of little thing that gets to you after a while."

Luella knew the difficulties of commercial migratory beekeeping. "Scheduling is impossible. Meals are hit-or-miss. The only vacations are when the bees are moved to different parts of the country. Working side by side twenty-four hours a day is a hard adjustment. But now we've gotten used to it. We probably wouldn't know what to do if we weren't together."

Horace, she said, fixes on things. Once he decides he wants something, that's it. "I was the only one he ever dated," she said. "He went out with me and he didn't want anyone else."

We stopped in Groveland, at David Miksa's queen-rearing farm. Miksa had four hundred queens for Luella, and he said he could hold a thousand more. We went to a queen house, a shed with a workroom and walk-in incubator. Miksa took a box of queen cells from the incubator and Luella began to count out her four hundred, setting them into egg crates and laying blankets of fiberglass insulation over them to keep the cells warm.

Because of the federal quarantine on the state of Florida, Miksa had lost most of his mail-order queen business. Tracheal mites, he worried, would put him out of business entirely. He wondered whether Florida didn't have the highest infestation of mites not because it had the most migratory beekeepers but because it had the best inspection system. "The thing to think about," Miksa said, "is that in two years we'll be wondering why we checked our hives so thoroughly"

Miksa's primary customers were the migratory and commercial beekeepers in Florida who were making increases for the move north. While we were there, a migratory beekeeper came into the shed for his order of queen cells. His name was Rufus Cox. He ran his bees along the East Coast, from Maine to Florida, doing pollination and making honey. Cox had on a visor cap, and his hair, a yellow gray, was tied back. His assistant was a muscled ex-Marine with a brush cut.

"How's Horace gonna get out of here?" Cox asked Luella, in a voice that was a smoker's deep rasp.

"Don't know." She kept counting.

Rufus took a drag from his cigarette. He had heard rumors. "If they lift the quarantine, Alabama and Georgia are all set," he said. "No bees going through. No bees at all. They're gonna burn hives. Gonna *burn hives.*"

"Gonna be able to move through South Carolina only at night," Miksa said. "And how can you claim mite-free? It's impossible. You'd have to sample every bee."

"I'm getting ready to stay up in Maine," Cox said. "I get my money from pollination and wild raspberry honey in the mountains of Maine. Hundred-and-twenty-pound average on that." He clapped his hands together. "Got it all sewed up. It's like orange down here. It's everywhere. Bears get on it, though. If the bears don't get it, the Fourth-of-July partiers do. They'll tear off a corner of my fence to watch the bears go in. I lost, what, a hundred hives up in Maine last year?" Cox's assistant nodded.

Suddenly Cox became testy. "You know what apple growers in Maine are offering? Forty, fifty dollars a hive, and it's going up. Blueberry may be going to fifty or sixty. We get two thousand hives up there." Cox punched his hands and grinned. "We just got to get up there."

Back in Deland, Horace called from the porch. He wanted a broom. He was in front of the incubator, looking hot. The queen cells in the incubator had begun to hatch out and the queens were crawling the walls. Horace had been brushing them to the floor.

The man that Bell called his closest friend was there, too. His name was Willard Ainsworth. He was taking queen cells from a carton and holding them to a lightbulb. Those he liked he was setting in an egg carton.

Horace was giving him a deal—half price, a dollar each.

Horace had met Ainsworth a few years back, when he pulled into a gas station one night with a truckload of hives. Ainsworth was pumping gas, and when he looked at the truck, he couldn't believe what he saw. He already had an interest in farming and husbandry. He grew hay, and kept a pig, and had a yard of gopher turtles. He even delivered his own children ("about eight head," Horace said). But Ainsworth had never imagined you could keep bees the way Horace Bell kept bees. Ainsworth offered to work for Bell for free, just to learn. Eventually Ainsworth got a semi and went on his own, but the two remained friends. Horace called Ainsworth a ball of fire.

Now the ball of fire was looking through wax cells to see how the occupants were doing, and the best beekeeper in Florida was sweeping queen bees into a pile on the floor, not liking it one bit.

17

"Whatchoo want to write a book for?" Wayne Knight said. He was one of Bell's foremen. He had been with Bell for thirteen years. His drawl, central Floridian, I gathered, made you remember that Florida was connected to the southern states.

And he was suspicious. I talked about the industry, but Knight thought I was on the trail of killer bees. "You heard of these African bays?" Knight asked. "These here *are* African bays." He knew a beekeeper who worked in South Africa, the origin of a strain of killer bees. That beekeeper told Knight they weren't hard to work. "No harder than these bays here." Knight picked a hive up in his arms and walked off.

The Bell crew had been splitting hives for weeks, and had made seven thousand hives into fourteen thousand. Horace had designed a splitting system using warehouses and stock hives. In a warehouse the crew worked along a column of hives. They opened a hive, removed half the frames, and placed them in an empty box, or hive body. These new hive bodies were then set on established, parent colonies, with a queen excluder between them, creating a pallet with six established, single-story hives underneath six new splits. During the next twenty-four hours, bees, attracted to the brood combs, would drift from the parent colony into the new split. The parent colony was thus both furnace and source of bees.

The rearranged pallets were moved to another warehouse. On the following day the upper boxes were removed, given bottom boards and covers, and transported to the field, where at dusk queen cells were introduced. It was a principle similar to the making of mating colonies in Reggie Wilbanks's mating yards, but on a larger scale, with full-sized units.

I had on a full bee suit. Everyone did. During the splitting operations bees flew from the hives. Because of the number of hives coming and going in the Bell yard there was always a great number of bees in the air. They drifted aimlessly, circled the warehouses, flew at the crews, landed on the trees. They were louder than the truck and forklift motors.

The bees had landed on the oaks near the warehouses and covered the leaves. On one oak, thirty feet up, was a cluster that looked like a wooden door. It was four feet across and six feet long, a solid body of bees festooned leg to leg that must have weighed thirty pounds. Pieces would break off and scatter in the air.

There was a hive body on the ridge of a warehouse roof over the entrance that must have been placed there for ornamentation, like a figurehead on a ship. Bees had filled it and hung from the entrance, like a spread of lava, flowing over the roof and down the wall. One wall of the warehouse was screened, and inside bees crawled the screen in perpetual motion, looking like water flowing upward. They bunched at the top and fell in heaps.

I helped split and carry hives for a while, but then I sat down on a hive to watch the work. Soon, though, my throat began to tickle and constrict, and I started coughing, One of the crew walked over to me. "It's the bees," he said. "All the dead bees make you cough." I stood up.

His name was Wayne Carter. They called him W-Two. Wayne Knight was W-One. Carter was from Blue Hill, Maine. A blueberry worker, he had met Andy Card during pollination season and subsequently left to work with him. Carter dug into a hive, took out a knife and cut out a chunk of honeycomb. "Want some?" he asked, holding out the knife. That was lunch.

W-Two's wife, Julie, was also working. Julie looked lost in her bee suit. It bagged at the legs and arms. Her hat slid down over her eyes. She looked very unhappy. Sometimes the bees crawling over her became too much to take. She'd freeze and call for Wayne, and he would brush the bees off, or shake her by the shoulders. Reactions like Julie's were one of the reasons Horace Bell paid some of the highest agricultural wages in Florida, or Maine, for that matter.

Another member of the crew was Gary Kelso, Nina's husband. He was nineteen, tall, slender, clownish, and a steady worker. When he started dating Nina, Horace hired him to fill in one day. He stayed on. He wanted to learn the bee business. "It's not bad today," he said, "because it's cooler. I got twenty stings on the back of the neck in five minutes yesterday. They get you through the veil, where it touches the skin." Gary mentioned Nina's pregnancy, and kicked a foot up. Now, he said, he wouldn't have to go to North Dakota.

At three o'clock four high school kids punched in, and with nine people working, the splitting operation quickened. By five o'clock three trucks were loaded and ready to go. We left Deland, drove to the St. Johns River, and followed a dirt road along a pasture to a stand of hives. I was in a truck with W-One and W-Two.

"These bees are mean," W-Two said.

"Reckon we can scatter them over yonder," said W-One. While we were bringing new hives in, Luella drove out, headed for Deland with four hundred hives that had been there long enough to build strength. They would go into orange groves the next day.

I put on a suit and helped take feeding jars off the departing hives.

Then Horace arrived, and started running the forklift. Heather Bell, Nina's sixteen-year-old sister, pulled off feeding jars and stacked hives, too. She was small, and couldn't have weighed much more than a hundred pounds, but she had a way of throwing her body weight so as to lunge about ten feet with a sixty- to seventy-pound hive. Recently Heather had become faster at the work. Horace had noticed this and given her at first a 10-cent and then a 25-cent raise.

"It's too hard sometimes," Heather said, "but I like it pretty much."

"Do you get stung often?" I could have come up with a better question. This was the one she probably got at school all the time.

"A bit," she said.

When the man who owned the pasture dropped by, Horace got off the forklift and talked to him. But the conversation didn't last long. His retriever dropped a Frisbee from its mouth, batted at its nose and chased its tail. Then the landowner started to pull bees from his hair, and ran to his truck. Horace, looking almost disappointed, went back to work.

When the sun went down, a breeze picked up and blew across the pasture. Two more truckloads of established hives were loaded up and ready to go, and all that remained was to arrange the splits and put queen cells in them. Horace and W-Two ran the forklifts. W-One climbed on a load and walked back over the hives, pulling a net. He flung straps over the netted load.

With the net covering the hives and darkness falling, Knight took off his hat and veil. His hair was red and curly, his sideburns thick and bootshaped. His nose was sharply cut, with a wide bridge.

He cinched down a strap and pulled on the ratchet clamp. "This one's twisted," he said. "That's all right. The bays won't mind."

Next strap. "I've been fired a hundred times," Knight said, "and threatened to quit almost as many."

Another strap, cinched down with a body lean. "Love it, hate it, the bee business. You can't stop thinking about it. That's what makes you hate it."

Another strap, and an economic theme: "You never know what beekeepers are making. They all lie to each other. You can't believe anything they say. It's all a secret, you know,"

And a favorite theme, the long workdays (this one was already ten hours long): "Once in a while we go home at six, but that ain't often."

Life on the road: "You ought to go up to Dakota. Dirt clay gravel roads up there, just like in the cowboy days. This town we were in, Philip, one restaurant, no 7-Elevens, six bars. A woman owned the only restaurant there. The service was *slow.* She'd sit there and look at us. We had to wait till she got ready. We called her Countdown. It got so we could sit there and count down—five, four, three, two, one—to when she was going to come over."

Knight flicked a bee off my sleeve. "A drought all summer up there. Grasshoppers, grasshoppers, everywhere, all over the clover. I was separated from my wife for three months one year. Last year, though, she came up. Stayed in a motel that was made out of a travel trailer. Nine hundred flies and a TV full of snow. We had two good weeks together."

Back to the favorite theme: "Crew of five guys up there. One got a divorce. His wife left him when he was gone. She didn't want him working bees, anyway. He come home too damn tired."

It was almost night and quite dark before the pallets of splits were set out in the field. They were arranged in clock-shaped patterns, twelve pallets of six hives each in a circle around a thirteenth pallet at the center, groups of seventy-eight hives.

Luella was back at the site with queen cells. She lifted the hive covers and pressed cells between the frames. Now and then bees sprayed out at her. She would place six cells in the six hives on a pallet, straighten up to stretch a stiff back, and move on to another pallet. A man, Jack Finlay, followed her with a flashlight.

The work was going slowly. The queens were just over the sixteen-day cusp. Many had hatched and then died trying to get through the bottom of the egg carton. Luella kept tossing dead cells away. She looked for activity, movement. "They don't bob their heads, they're gone," she said. Sometimes she tossed five away before she found a good one. "There goes a dollar, and another," she'd say.

When Luella ran out of queen cells, she would run back to the van for another carton. Finlay held the light on the van when she did this. "You seen her work before?" he said. "She's the workingest woman I ever did see. She used to drive that forklift around the plant in South Dakota. I never saw anything like that. She could move some barrels."

Finlay had joined Bell in North Dakota. "I drove the honey in for them. I'm a truck driver. I walked into the restaurant one morning and Horace was there with this man from town. He says to Horace, 'There's a truck driver, get his ass out of town.' So I started driving honey from sunflower locations in North Dakota down to the extracting plant in South Dakota. Then I helped move the bees down here. I was supposed to be working on the trucks, but Horace put me on the bees. I been doing more of that than anything else. I caught a sting in the earlobe the other day. I didn't like that at all. You ever get stung in the ear? You ought to get one there to see what it's like. Put that in a book."

Luella came striding by with another egg crate of queen cells. Jack Finlay lit her way and smiled out of one side of his mouth. "I'm supposed to be a truck driver, you know."

I left with Wayne Knight, and as we drove away I looked back. All the light had drained from the sky, and there was only the shaky beam of light from Jack Finlay to Luella's hand.

18

The three trucks with the twelve hundred hives from the St. Johns River were to go into orange groves in Fort Pierce. Horace was going to drive one truck, Nina the second, and Gary the third, with the forklift trailer. I waited all morning around the house, in the kitchen, talking with the housekeeper (who was from Michigan and whose husband was working for Horace to learn to keep bees commercially) and waiting for Horace to call me. "You better run," she said. "He's leaving." I heard an engine racing.

When I jumped into the truck, Horace smiled, just a little lengthening of the lip, and I thought of it as encouraging. I had heard that Horace didn't talk to people. He roared off and the rest of the caravan started up behind.

We passed by a withered orange grove. "Are they going to replant these groves?"

"No," Horace said. "They'll plant Yankees on them. They're probably building a house for you right now."

It was three hours from Deland to Fort Pierce. I watched the landscape. We talked about barge beekeeping in the tupelo swamps of the panhandle, and about a packing company in Edgewater that sold specialty honeys.

"When I was five years old, my grandfather gave us five hives," Horace said, loudly. I could feel his voice vibrating through my fingernails. "He had a hundred fifty hives, a big outfit for that time. I kind of took the hives over. I went to school and sold the honey for twenty-five cents a jar. It went like hotcakes. We just squeezed it out of the combs by hand. Kids would go back and ask their mama for a quarter so they could buy a jar of honey."

It was a country school, Horace said, and he didn't wear shoes to class until he was in the fifth grade. Horace liked to take to the woods, exploring the territory along the St. Johns River. At night he took a flashlight, sat in trees, and stared at anything that moved. He liked to watch ant colonies managing aphids. He loved to play in an abandoned mansion that a Confederate general had supposedly hidden out in. He still plays on the St. Johns River now and then. When he wants to entertain, he takes people out on the river to watch alligators and manatees.

"I was always interested in agriculture," Horace said. "I had pigs and chickens, all that stuff, but the market was all screwed up on them. Once I got into the honey business I found out that it didn't matter if you had five hundred pounds of honey or five thousand. You just got it all on a truck and then they'd buy it. They weren't using drums yet. I used milk cans.

"There was this old man who gave me some experience. He was from Georgia, and he was poor all his life. He couldn't afford bees when he was a kid, but he was so intrigued by bees that he dug up yellow jackets and tried to hive them. He came down here, and he raised queens that were famous all over the country, queens raised for this area.

"He had some skills I haven't got yet, stuff I have to look forward to. One day he said, 'Look at that hive. I haven't touched it in forty-seven years.' That's something I would do, watch one hive for so many years.

He'd breed for certain traits. He was as religious as anyone, and you know, he didn't exaggerate. What this old man said was, get bees for this area.

"I kind of studied under him, a protégé. He always got on my ass because I was so gung ho in the bee business. I would never relax. I never believed that he was right until three or four years ago."

We moved south on I-95, passing by burned-out orange groves. Now and then Horace called Nina on the CB and asked how she was doing.

In 1962, Horace said, when he was seventeen, he heard that the Farmer's Home Administration would lend money to beekeepers. They told him to come back when he was twenty-one. So Horace went to a lawyer who told him to have his age disability removed. Horace went back to the FHA office ("They figured if I was that gung ho, they ought to give me the loan"), borrowed $4,000 and invested in beekeeping. Two years later he borrowed another $10,000 and bought more bees. He migrated to Ohio with seven thousand hives. It was an extraordinary migration, especially for someone only nineteen years old, with a crew loading and unloading by hand.

"I got married when I was eighteen. When Luella got pregnant, I got my draft notice. I told this to one beekeeper who I'd bought out. He had been in World War Two and he told me that beekeeping was considered an essential industry. The reason was that back in World War Two there was a shortage of sugar, and they used beeswax on the shells and all. I went to the draft board and told them that. They finally gave it to me. I was probably the last one to get out of the draft for beekeeping.

"Times were pretty bad in the early going. Before the price of honey skyrocketed in the late sixties we used to go out and kill rabbits, hunt things to get something to eat. I was up to seven thousand hives, back in 1968, 1969, when we had a couple of bad crops, and by 1970, 1971 we'd dropped back to two or three thousand hives. Then when the price went up, we had the equipment, so we went back to that number.

"Anytime you get a situation like the one we were in, somebody without any skill can get by, working ten to twenty hours a day. We were like that when we started out, working fourteen hours a day. You can overcome the low return by working the hours."

In the bad years Horace nearly went bankrupt, and couldn't pay his loans. Beekeeping was still in a postwar decline. Honey prices had gone from a high of 24 cents a pound down to 10 cents. "The lowest price I ever got for honey was nine cents a pound," Horace said, "back in the fifties. Then in the late sixties, into the early seventies, things changed, and the price went way up again. They attributed it to the health food boom, to all the damn hippies eating honey."

With all the damn hippies factored into the honey business, prices jumped from 10 to 50 cents a pound. "For a couple of years everything stayed the same, queens, foundation, wood. Everything didn't catch up right off. A queen at ten cents a pound takes ten pounds of honey. It's all according to the price of honey, five-dollar queens or dollar queens. So for two years we had five times the amount of money to invest. When the price went up, we sold the lowest quality of honey for top dollar."

Horace got out of debt, promising himself he would never get so close to bankruptcy again. From then on, he paid cash, for his house, his cars, trucks. If he wanted a new car, he would pass on a load of corn syrup and use the money to buy the car.

"The real changes in commercial beekeeping were from 1972 to 1980. All through those years people came out of the woodwork. Every Tom, Dick, and Harry with a hive of bees jumped into beekeeping. It's kind of attractive you know, beekeeping. On paper it looks like a damn gold mine. It isn't, really, but for people like us it was a damn gold mine."

19

In 1985 the United States government was holding one-half of the world's honey stocks, and it had become the world's greatest honey buyer. This honey was stored in warehouses, by beekeepers, for a fee. Nearly all commercial beekeepers were using the federal honey support program, and though it would all change after the 1985 season, the gold mine Horace Bell spoke of had raised a few eyebrows, in Congress and in the national media. It was hard for people to accept the idea of a beekeeper with a gross income over a million dollars, and nearly impossible to accept the idea of that beekeeper turning his entire crop over to the government.

The honey support program went back to the period following World War Two, when honey prices dropped to 9 cents a pound. Commercial beekeepers lobbied Congress for support. They argued that a healthy beekeeping industry was essential for American agriculture, and they were right. So honey was included in the Agricultural Act of 1949, which was plugged into a support program that protected such commodities as wheat, corn, and rice. As of the 1950 season honey prices were supported by parity.

The parity formula is a system of averaging outlined by Congress. It reflects relative value, first according to the buying power of the base years between 1910 and 1914, and additionally according to the most recent ten years of market prices.

In 1950 the parity formula indicated a honey price of 15 cents a pound. Since support prices were at 60 percent of parity, the honey support price in 1950 was 9 cents a pound. During the next twenty years prices didn't change much. The support price was 8.6 cents in 1960, and 13 cents in 1970.

A beekeeper used the honey support program by taking a loan and using his honey as collateral. The loan allowed him to pay operating costs, and to hold the honey until he found the best market. After the sale he paid off the loan. But the beekeeper could default on the loan and forfeit the honey to the federal government. This rarely happened in the first two decades of the honey support program, because the support price was lower than the market price.

Then environmental and nutritional awareness changed. It was just as Horace said: All the damn hippies started eating honey. Refined sugar was linked with a number of health problems. Honey, on the other hand, though still a sugar, had enzymes, trace minerals, traces of vitamins, and a mystique of beneficent health. Between 1968 and 1972 per-capita consumption of honey increased in the United States. It was a small amount per citizen, only a few tablespoons, or .012 pound. But this worked out to a rise in national consumption from 215 million to 252 million pounds per year.

"Hell," Horace told me, "all people have to do is eat a damn tablespoon of honey every day and we wouldn't be able to produce enough of it."

New, higher prices entered into the parity formula, and by 1984 the support price was 65.8 cents a pound.

American beekeepers have never produced enough honey to meet American demand. When demand increased, Mexico and Argentina increased production.

Foreign prices dropped as parity prices continued to climb—an upward curve crossed a downward curve. There were two ways of expressing this change. As the beekeepers saw it, federal support prices rose above wholesale market prices. As some politicians saw it, market prices dropped below the federal support price, and the honey support program had become a waste of taxpayer dollars.

In 1984 beekeepers forfeited 128 million pounds of honey, almost two-thirds of the U.S. honey crop. The government was holding 165 million pounds of honey, roughly half the world stocks. Much of the honey was stored by commercial beekeepers, who had the warehouses and equipment at hand. They received a fee of 10 cents a hundredweight. It was the least expensive way to handle such a supply.

Foreign prices, meanwhile, remained absurdly low. Mexican honey, for example, was 35.8 cents a pound in 1983, and Mexico was exporting 90 percent of its crop. Because the forfeited American honey was not entering the domestic market—it was used in various government food programs—imports increased. In 1984 the import figure and the forfeit figure were the same—128 million pounds.

Horace Bell saw the government program for what it was—merely a tool for the beekeeper. He used it "as much as anybody, maybe more." Horace took a loan on 360,000 pounds in 1967 and did not default on it. In 1983, however, Horace forfeited a million pounds. He reasoned it was a good thing, His forfeiture cleared the market for other beekeepers.

Horace wanted the government to tax imported honey up to 50 percent of its value. In 1976, in fact, the International Trade Commission recommended that President Carter establish import relief for beekeepers. But President Carter decided against it.

It seemed likely in 1985 that the honey support program would be terminated. The beekeepers, competing with low prices on foreign honey, feared for their livelihoods.

20

For many years Glenn Gibson traveled from Oklahoma to Washington, D.C., to represent commercial beekeepers. Gibson was the honey lobbyist.

He was also a beekeeper, though he was more of a honey packer and business manager than someone who worked in the field. His company, Clover Bloom Honey, ran a thousand to thirteen hundred hives, bought honey from other operations, and distributed as much as a million pounds of honey a year to markets in Oklahoma.

"I was born in Oklahoma, April 8, 1917," Gibson told me. "We were sharecroppers, raised cotton and feed grains and some wheat, near Hollis, the county seat. I chopped cotton, milked cows, ran a one-row cultivator. That was in the day when you were looking those mortgaged mules in the hind end. We'd hay, too, with about thirty people running the hay baler. It looked like a herd of livestock working a forty-acre field when you were baling hay. The last day I worked at a hay baler, now this may sound like a lie, but it's the truth. No money in the country, 1934, five bales of hay, for a day's work. Now that day, and I remember that very well, was twelve hours. And if I could have sold that hay, I could have got fifteen cents a bale for it. That was hard times."

Gibson entered the honey industry in 1935, while attending Hills Business College in Oklahoma City, where he studied accounting. He took a part-time job at the Clover Bloom Honey Company, working three hours a day five days a week and five hours on Saturday, for $5 a week. He became a manager, and five years after joining the operation, he put down $500 and became a partner, and ultimately the sole owner.

Gibson moved Clover Bloom from Oklahoma City to Minco, which was central to the company's nectar sources. Politics, and political power, was a source of fascination for Gibson. He wanted to serve and hold office. He was the mayor of Minco, in 1978. And he was the only person to hold all four executive offices in both the American Beekeeping Federation and the American Honey Producers. He helped form the AHP, which exists solely for political action.

Gibson made his first trip to Washington in 1951, when he was thirty-four. Over the years he had learned a few things, and described it this way: "When someone's trying to cut my throat, I know when they're doing it. Most of the time."

He also knew the charm of honey. Before trips to Washington he sent cases of honey to the people he planned to call upon. "Try to get a guy's attention for six months, and you give him some honey, you got him."

In 1962 Gibson was a Republican candidate for Congress in Oklahoma's sixth district, and got, he said, 48 percent of the vote. It was the highlight of his life, coming so close to winning a congressional seat.

But he went to Washington nevertheless, to work not for the people of Oklahoma but for the beekeeper. He averaged five trips per year, through the sixties and seventies. In the mid-eighties, after he had turned Clover Bloom over to his daughter and son-in-law, and with the uncertain future of the honey support program, Gibson averaged a trip a month.

During these years he helped obtain funding for the insecticide indemnity program, insecticide research at the USDA lab in Wyoming, and the honeybee stock center in Baton Rouge.

His energy was formidable, but his health was poor. He was a heavy smoker, and hacked incessantly. At the end of our day of trooping, with twenty or more visits behind us, and three to go, Gibson became impatient with an elevator, and turned for the stairs, but had to use both hands on the railing to pull himself up the steps.

His purposes for the incessant trips were partly selfish, Gibson said. Favorable legislation would help Clover Bloom. Another purpose was foolish. "The job isn't complete and never will be. I would like to train someone to follow it up. I'm anxious to see the producer's welfare looked after.

"Washington is an exciting game," Gibson said. "When you can rub elbows with the power of the country, as friends, swap jokes, this thing, that's really top stuff in my book."

It was morning. A white mist hung over the Potomac. There were four people in the taxi. Crossing a bridge from Arlington to Washington, Gibson noticed a tree with branches hanging over the railing. It was ripe with small white fruit. "What feeds on that?" he asked. "Those berries. Possums? Woodchucks?"

"I don't know," Robert Cook said, after a puzzled look. Cook had at one time administered the pesticide indemnity program for the USDA. He was along to guide Gibson through the passageways and cellarways of Congress.

Gibson gave an order to a beekeeper sitting in the front seat of the taxi. "Jack, tell that driver to go to the Hart Building."

"I did," Jack said. Jack Meyer, Junior, was a fourth-generation beekeeper, and forty-three years old. His family's operation produced a million pounds of honey a year. Meyer had come along to provide some power of constituency. Since Meyer was from South Dakota, with winter operations in Louisiana, he could work senators and representatives from those states. Spadework, Gibson called it.

The party stopped at the Hart Senate Office Building. Gibson got out of the taxi, took his Kool from between his teeth, and said, "I feel like we have been thoroughly mistreated with the information coming out of the Department of Agriculture. Particularly the GAO report."

A report had recently come out of the General Accounting Office, titled "Federal Price Support for Honey Should Be Phased Out." The GAO staff had interviewed people in the beekeeping industry, and come up with three conclusions: 1) support payments were so high that beekeepers were making honey instead of pollinating crops; 2) crop producers would pay higher pollination fees if necessary; 3) only 1 or 2 percent of all the beekeepers in the United States used the honey support program.

Several members of Congress had praised the report, but Gibson thought the conclusions had come first and the evidence second, that the report had been written to develop a set of arguments against the loan program. Gibson's spadework, he said, would be to counter damage done by the GAO report.

"Silvio Conte used the damn thing, and Mr. Quayle, too," Gibson said. "We've got five or six thousand beekeepers who produce honey that make up the market. By no stretch of the imagination can two hundred thousand hobbyists affect the price of honey either way. Or use the program, so why the hell did they include them in the report and make big deals about the percentage of the people participating in the program? Man, we're just not getting the truth on all this rhetoric."

Gibson wanted to start the day with a visit to Senator Dan Quayle. Quayle had sponsored an amendment to discontinue the honey support program. Using the GAO report he had made a speech in the Senate with these lines:

"This money is not for the family farm out there," Quayle had said. "This is for the millionaire. You have a millionaire bees club around America.

"They probably all contacted everybody. They are very wealthy, influential people. They own a lot of bees. They want to have more bees. They've got bees everywhere. You know, they buzz around here. Sometimes there are not even bees buzzing around here. Other things buzz around here.

"But bees? They are important. We are probably going to hear some arguments about bees being important to our national security. We are not going to be able to get along without the bees, particularly the hundred million dollars' worth of bees—buzz, buzz, buzz; bees, bees, bees. They are buzzing about everywhere. We are going to have a lot of bees on this floor and bees around everywhere, and bees for everybody—one hundred million dollars' worth for the millionaire—not the little guy, but for the millionaire."

Quayle went on. "This is a lesson of political science. How to get organized, how to get organized at the grassroots level and make sure that you contact that senator and that congressman, and make sure they preserve the status quo. Make sure they preserve that sweet subsidy that has been ripping off the taxpayers for a number of years. Make sure that they are able to organize, and will come up and, by golly, if you do not vote for them, watch out for those bees, watch out for those bees. We need bees. We like the little buzz and hum. We like bees."

So Gibson wanted to do a little spadework with Quayle. He walked into Quayle's office, the rest of the party at his heels, and announced himself "Glenn Gibson, ma'am, President, American Honey Producers. We'd like to talk to Senator Quayle, or one of his ag aides, please." The secretary made a call, but the senator was in a meeting, and so was his agricultural aide. Gibson said he'd try again later.

Outside Quayle's office Gibson took out a Kool, lit it with shaking hands, and walked on. If Gibson was wise, he was also wizened. He wore a gray suit that contrasted nicely with the silver tones in his brush-cut hair. He was always in and out of his pockets for Kools and matches, and the flaps were often dogeared. Now he pulled out a name tag, and became annoyed that he hadn't pinned it on earlier. The tag read, "Glenn Gibson, President, AHP."

Gibson, Meyer, and Cook then went to visit Senator Dan Nickles, of Oklahoma. Gibson was in home territory now, and his ease showed. The decor in Nickles's office was a mix of college football and American Indian. There were three gleaming football helmets—Oklahoma, Oklahoma State, Tulsa—on a shelf in the middle of the wall. On shelves and tables were books about cowboys, an Indian headdress, a tom-tom. The secretary offered coffee.

The senator was not available, but Les Brorscen, an agricultural aide, appeared and led us to a conference room. Brorscen explained the current situation with the honey program. The House had passed a proposal to put a $250,000 cap on support payments. The Senate had come up with a marketing loan program—the buyback program that would replace the honey support program and put American beekeepers back into the marketplace, giving them payments that were the difference between wholesale and loan rate.

Gibson didn't like either idea. The ideal solution would be import protection; barring that, the present program should continue. Gibson told Brorscen that he doubted there were twenty senators in Congress with an understanding of the bee business.

"Now when you mention wheat you've got a vision of combines, but when you mention beekeeping, most of the people you talk to—and this was true in my campaign back there in 1962 for Congress—they visualize that old man up the end of the block that had three hives in the backyard. I remember very well those sons-of-a-bitches over there in Channel Seven, they'd mention my name and chuckle. They couldn't resist because it seemed it was a damn joke for a beekeeper to be running. A beekeeper couldn't be anybody but an old man that was half kooky. And a lot of that prevails here. You know, you get it in some of the conversation with the dumb questions that are asked. Now every damn guy for the last thirty years that I've contacted up here, Les, that understands our bee business, has been a friendly congressman."

"The question," Brorscen said, "is whether beekeeping is crucial to the ecosystem of Oklahoma."

Jack Meyer had been silent but now he broke in. "Back to this loan cap again. We're basically locked into honey production. It isn't like the farmer who can get into different programs, like wheat or corn or whatever. With honey, we're locked right in. With some of us like myself it's four generations of business. We've tried to expand and grow, and this cap would really affect us. Our livelihood's on the line."

A door opened, and the secretary said Senator Nickles was ready. We all filed from the room. Nickles was standing by his desk. He smiled, though guardedly, and said he had to leave soon to vote. "We have to do this quickly," he said.

Nickles, Gibson, Cook, Meyer, and I stood in a line. A photographer said, "Talk." Nickles asked Gibson how long he was in town for, but Gibson, who had fixed a smile on his face, and wasn't about to drop it, didn't answer.

We returned to Senator Quayle's office, but no one was available. Next the party walked to the office of California Representative Tony Coelho, chairman of the agricultural committee on livestock, dairy, and poultry, and both Gibson and Meyer argued with an aide against the proposed cap on support payments. We stopped in the office of Senator Melcher of Montana and Gibson left his card with Melcher, "a friend of the beekeeper."

At lunch at the Senate dining room, a congressional aide told Gibson that both pasta and shoes had just received import protection. Gibson shook his head in dismay.

1:00. The office of Representative Jim Lightfoot of Iowa, to make plans for a hearing on the loan program in Sioux City. Lightfoot said that when he was a radio journalist he worked to restrict the use of pesticides around apiaries.

2:10. The office of Senator Thad Cochran of Mississippi, another friend of the beekeeper, where Gibson spoke with an aide about the necessity of import protection.

3:00. The office of Steve Adams, a staff consultant on the House Agricultural Committee, where Gibson presented his arguments for the continuation of the loan program. "It's a can of worms. If the House proposal for a loan cap prevails we got a hell of a lot of trouble. Because biggies like Jack Meyer will come into the market and cut my throat. Isn't that right, Jack?"

Meyer smiled. "You bet."

3:50. The office of Representative Manuel Lujan, New Mexico, another friend of the beekeeper.

4:15. Iowa Senator Berkeley Bedell's office, for talk with an aide.

4:35. North Dakota Representative Barry Dorgan's office. To an aide, Gibson said, "Guys like Meyer here will come down and cut my throat. My opinion is, that wherever a congressman understands our business, we've got a friend in Congress."

4:55. The office of Representative Glenn English, Oklahoma. Gibson told a secretary he wanted to chicken-fry the boys from the Department of Agriculture, those who had circulated a mock application for a honey support loan, signed by "Plenty Rich."

5:00. The office of Representative Silvio Conte, of Massachusetts, the archenemy of the beekeeper. Gibson was hoping to get some spadework out of me, since Conte is my representative. "Work some sense into those people," Gibson said. He and the rest of the party waited around a corner while I visited with my rep.

Conte's office was the busiest I'd seen. The entryway wall was nearly covered with photographs of the congressman with famous people. I told the receptionist I was from Massachusetts and wanted a copy of the speech about discontinuation of the loan program.

An aide led me into an office where he looked among many piles of paper and in many cabinets and drawers before he found the speech. He interrupted a red-eyed speech writer. "Where's the speech? You know, the one he had to stay up all night fixing?" The writer glared at the aide and at me and said he didn't know where it was.

But then the aide found it, and made a copy. He said it was outrageous that the government paid a beekeeper a million dollars for honey. I said that beekeepers were trapped into the program. He said the Great Depression was caused by protectionist policies, that we would all have to get used to a lower standard of living. I said hobbyist beekeepers would not pollinate crops.

And so went the spadework. I left with the Conte speech. The first three paragraphs went like this:

"Mr. Chairman, a tiny, little bee lit on my shoulder on the way back from the floor last week when my amendment to strike the honey program went down on a point of order. This little bee buzzed in my ear and said, 'Don't give up, Silvio—you stick right in there because the good name of bees is being besmirched all over this great country—especially in North Dakota where one producer got $1 million in support payments.' The little bee told me that Congressman Frank would try to stop this super-sweet deal by placing a limit of $250,000 on the amount of payments to the big bees in the honeycomb business. There is no limit today. My amendment goes Mr. Frank one better and does away with the entire program. This little bee had all the facts. He told me that the nonrecourse loan rate for honey is at over 60 cents while the price we pay for imported honey is at 40 cents, so we are importing more and more honey.

"The GAO has looked at this program and they say we are getting stung. We are spending nearly $100 million to store 100 million pounds of honey each year in order to benefit only 2,500 commercial producers.

At the same time we are importing 100 million pounds to put on grocery shelves. You'd have to have bees in your bonnet not to see how far out of line this situation is.

"I think everyone gets stung once or twice in their life, but this country is getting stung over and over again. Now the bees are getting into the act. Why just last week in Hollywood, Florida, when the bees found out about how the Americans are being robbed by this program they got out of their keeper's truck and refused to pay the toll on the Florida turnpike. I tell you these little fellers are in revolt over the way we raise the price on their diligently created product. Help me restore some dignity to the little bee and let's swat this program once and for all."

We walked outside and along the steps in front of the Capitol building. The day was full of soft light, on the buildings, the maple trees, the grass, the zinnias.

"What are you going to do next?" I asked Glenn Gibson. He was walking with a long-legged stride, looking down at the sidewalk, fists deep in his pockets, a Kool between his teeth.

"Start over," Gibson said.

I turned to look at the Capitol, at the towering dome, the people streaming from the entrance. It was, I realized, modeled on a beehive, and I was, I realized, seeing the world in apicultural imagery.

21

Ultimately the loan program was altered, not discontinued. The 1985 farm bill added a marketing loan system to the honey support program. Beekeepers call it the buyback program.

The idea was to induce American beekeepers to sell their honey on the wholesale market, after which they would receive a differential figure between wholesale and market prices.

In 1986 the average loan rate for honey was 64 cents a pound. A beekeeper could take a loan at 64 cents, and then pay it back at a lower rate determined by the USDA on a monthly basis. The payback rate began at 52 cents in July 1986 and subsequently dropped to 44 cents, where it remained for many months, and then dropped again. A beekeeper bought back 64-cent honey, sold it on the open market, and repaid the government 44 cents. The 44-cent figure was slightly below world market levels, so the beekeeper could compete with foreign producers and make a profit. If he used the buyback program, he paid no interest on the loans.

During the first two months of the buyback program, honey imports dropped by 10 million pounds, and there was a record use of the honey support program by American beekeepers. Honey cooperatives, Sue Bee chief among them, used the buyback provision, and in 1986 more than 150 million pounds of honey went on loan.

There was a limit of $250,000 on the payback (the difference between the higher loan rate and the lower payback rate, between 64 and 44 cents, for example). At a 20-cent payback, it would take 1,250,000 pounds of honey to reach that limit.

In 1990 a beekeeper would have to produce 2,265,000 pounds of honey to exhaust the resources of the honey support program. Only Horace Bell and a few other beekeepers were capable of producing such a honey crop.

22

"Orange for ten miles in any direction," Horace said. We were well into orange country, with groves in full bloom on both sides of the highway. The fragrance of citrus blossoms blew in the windows. Jeff Kalmes had once told me that he didn't like working in orange groves, that the scent was inescapable, like a woman with too much perfume. Horace didn't feel that way, especially after the freeze. "That's what I miss, that smell coming in my bedroom window."

It was hard to believe that only an hour's drive north of Vero Beach, the citrus industry had been devastated. And Florida was still the top citrus producer, harvesting seven million tons of fruit a year, two thirds of the national crop. Even though 180,000 acres of citrus trees had been lost to freezes, another 570,000 remained.

Nearly six hundred thousand hives were moved into orange groves each year. When we drove by viable orchards, we drove through streams of bees, and there was no way to avoid the ruptures and spots of nectar on the windshield. "You should see it in North Dakota," Horace said. "It's nothing but a dry desert up there. When they hit, it's about as thick as honey."

Horace had said that his practice was to feed corn syrup and take the honey. "I was the first one to feed corn syrup in Florida," Horace said. "I was the first one to buy a five-thousand-gallon tank and buy it by the truckload. Everyone was against it. They watched me and they checked me, but I had figured things out. For two or three years I was the only one feeding with corn syrup. Then I started selling it to other beekeepers.

"We know that corn syrup increases population. You can, with feeding, make bees think it's spring. A bee can store honey, somehow they evolved to store honey, from living in a cold climate. They had to learn how to invert complex sugar to simple sugar so they could store it. High-fructose corn syrup is levulose and dextrose. Honey is, too. When the bee inverts the honey, altering it by means of an enzyme, something else says, 'Hey, it's time to raise more bees.' I think you can do that by feeding corn syrup.

"One thing I can say is I never adulterated honey with corn syrup. One packer got rich on it. He could buy corn syrup for six cents a pound. When honey went up to fifty-six cents a pound, he got rich. I just took a different angle. Bees eat a certain amount of honey each year. Don't let them eat a drop of honey. Let them eat corn syrup. So I feed my bees corn syrup instead of honey."

In Fort Pierce we turned off I-95 and drove along the roads through the orange groves. Nina and Gary were close behind. I wanted to ask Horace a question before we stopped. How many hives did he own? Beekeepers are often hesitant about answering this question, sometimes because they don't know, but really because it's a private matter, too, like other questions of ownership.

"I wouldn't call them all a hive today, but they might be a hive next month. Right now I think we got about twenty-five thousand. We're planning on leaving Florida with thirty thousand." Only two other beekeeping companies approached that count, Powers Apiaries, with twenty-nine thousand hives, and Richard Adee, in South Dakota, who had about thirty thousand. Horace was making a risky move, with the freeze, the quarantine, the state of the support program. Andy Card described Horace's adventurousness this way: "Horace likes to flip the last dime on the table and wait to see what he comes up with."

We stopped at a store, and Horace called a grower. Then we drove to a waterway at the edge of a grove and turned down a sand road. Horace stopped and inspected a bridge over a culvert to see if it was wide enough for the trucks. There was about two feet to spare.

The grove owner came, and Horace left in his car. Nina was asleep. I picked a grapefruit and walked into the grove. The trees were both in bloom and heavy with fruit, and humming with bees. Horace's twelve hundred hives would surely increase the volume.

Nina was awake when I walked back. If she had looked slightly pregnant the day before, she looked hugely pregnant now, with her bee suit unzipped. The steering wheel made for a tight fit in the driver's seat. "Long ride," she said. "Those bumps are hard." Gary got into the truck, and they shared a lunch.

Horace returned and led the way into the orange grove. The trucks leaped and waddled, the engines whined, and it took Gary, pulling the forklift trailer, three tries to get by the culvert. His tires were inches from the bank.

The plan was to make three apiaries of four hundred hives each. We stopped half a mile in. It was a good spot for bees, not only because of the nectar flow, but because there was a water source, too. Bees need water to lower hive temperatures. When it gets too hot, they set drops of water on the combs and hold flexed sheets of water in their mouths while house bees line up and fan air currents. When water is needed, house bees will refuse to accept nectar, but enthusiastically accept water.

We put bee suits on. Gary unfastened the straps on Horace's truck, and then he pulled back the nets. A cloud rose from the truck and expanded. Part of it swarmed at me, searching for a way into my suit.

Maybe Wayne Knight was right, maybe these were African bays. Since the cuffs on the suit I was wearing were too high, I walked with a slight crouch to keep the pant legs low.

"Not gonna smoke these bees?" Nina asked Gary. Often, Horace didn't bother. They thought about it for a moment, then lit the smokers.

Horace started the forklift and dug into the load, while Nina folded the nets and Gary rolled up the ratchet straps. Horace set the pallets down in diagonal lines. He had said that the bees would smell the orange nectar and fly out of the hives for it before taking a careful look at the surrounding landmarks. They would come back with a full load and not know where to go. Inevitably they would end up at the last hive in a row, and overcrowd it. The irregular patterns, he said, made orientation and memorization easier.

I walked around like a nucleus with electrons, offered to help, rolled some straps, and then stayed out of the way. Nina climbed on the truck bed, knelt down, puffed the smoker bellows, but was out of fuel. Gary popped up with a handful of pine needles, smiling.

A car went by, the grove owner, and fifty yards up the road he turned and stopped. He got out of the car, walked to the bank of the waterway, turned his back and went about adding to the supply. But immediately he grabbed the back of his neck, and pulled at his hair. Horace calmly got off the forklift, got into the car again, and they left. Nina took over the forklift.

"He's different, huh?" Gary said. "Strange guy. Works you like a son of a bitch. It's O.K., though, because he pays you for it. The guys who have been working for a long time, they make good money. That's why I want to learn it, so I can make it, too." He slapped down the lid of a smoker. "I ain't seen it yet, though."

When Nina finished the load—it took an hour—we moved half a mile farther down the waterway, and she unloaded another truck. An hour later, four hundred hives down, we moved again, to the far corner of the grove, up against a row of tall palm trees. The sun was getting low, dusty and red, as it touched down over the orange treetops.

While Nina lifted the pallets from the truck, Gary kept ahead of her, puffing smoke into the hive entrances. They worked well together. When she gave just a flick of her wrist, Gary jumped off, got in the cab, and backed up, so that she would have more room to turn the forklift.

There were two hundred hives to go. The sun grew larger and redder as it dropped behind the dusty, smoky air. The bright leaves of the orange trees, the water in the canal, even the trunks of the palms took on the red light. The folds in Nina's bee suit flashed red as she spun from the truck. The sides of the hives were red. The bees—the air was so full of bees—were copper red in the air, a cloud of red and copper vibrating bees, into which Nina turned, stirring up more copper-red dust, for a few brief minutes, until it was settled and dusky and cool, contracted and quiet.

23

We were traveling north again, toward Deland. Horace's hair stuck up in tufts, from wearing the bee helmet. He was hoarse from shouting over the noise of the engine. He had begun to talk about some of his more fanciful ideas. For instance, Horace wanted to genetically encode his bees, for more accurate identification. He wanted to attach magnets to his queen bees, and when it was time to requeen, he could just hit the controls, zap them up, and pinch their damn heads off. When a city passed a zoning ordinance against bees, Horace placed his hives five miles away from the black mangrove swamps in that city, and found that he got just as much honey. The bees, he speculated, must have started evaporating the nectar during the flight home, so saving the house bees some of their work and thereby freeing them to go out foraging, too.

"Another thing," Horace said. "That place we set twelve hundred hives. We could take twelve thousand there. This old man I was talking about, who raised queens, he calls it multicolony outreach. He's done some thinking about it. Mostly by hisself. He had a lot of time to think."

And Horace had his own ideas about the effectiveness of bee communication. "You couldn't find a more gregarious insect than a bee. They're more gregarious in the field. I'm the one come up with this. When the area is saturated with bees it's like a school of fish or birds of a feather flock together. With one hive they're going to get out there. You get out among the bees you'll see. They kind of hang together. Big bee yards they go out farther, to get to the flowers.

"I am a little worried about the hives we just left. Each year I spend quite a little money on liability insurance, because you never know when somebody's going to get stung to death. But we've found that means bees make more honey. Now there's people in South America running bees like we do that's got killer bees. Possibly they are a little more aggressive, but we can work them. Actually I am a little concerned. But maybe they will make twice as much honey.

"Somebody told me they had bees that are so gentle they didn't even fly out of the hive. They walked out. They were so lazy they wouldn't get anything. An African bee wants to be the first son of a bitch to get out there and get it."

Horace had decided that we would stop in a town called Mims and load five hundred hives. Gary and Nina did not look happy, but they did not complain. Nina had been at work fifteen hours. And there wouldn't be much rest that night. The rule was that work started at seven, regardless of when they got in the night before.

They parked the trucks in a field of cabbage palms. The hives were beyond this field, in a grove of pines. It was a good location, well protected. Finding spots like this one was difficult, and Horace had to find dozens of them. It took time, because you not only had to find the land, you had to locate the landowner, too, and then secure permission. Sometimes Horace just didn't have time for this, and he wasn't the chatty type anyway. Some beekeepers, Horace among them, have looser senses of landownership. Beekeeping, after all, is like fishing, in that the flowers, like the fish, are out there for the taking. Placement is an impediment.

They are called drops, these abbreviated, anonymous placements. Bees, of course, can be a nuisance. There was a time when Horace's crew placed a large apiary near a zoo. When the nectar flow stopped, the bees, following the scent of soft drinks, moved over to the zoo's snack bar. Bell hives have been gassed by offended landowners. In Sanford, a town near Deland, trenches had been cut across roads to keep Bell bee trucks out.

It was night. No one would see us there. Horace drove the forklift, and for the first hour he just moved pallets around. Gary carried hives by hand and filled out pallets. Nina worked the smoker. It was hot, it was sweaty, it was hungry, it was tired, it was uncomfortable, with bees crawling up the sleeve, with the random sting, with the smell of pine smoke that had become noisome. It was six months pregnant.

But it was something else, too. It was a clear sky full of stars, and a warm, pleasant March night. It was the scent of nectar drifting from the hives, and the effervescent sound of the bees crowding at the hive entrances. It was a productive day.

Nina picked pine needles from the ground and stuffed them into the smoker, keeping a cloud of smoke ahead of Horace and the forklift. At eleven-thirty the loading was done. Gary fastened ratchet straps down, and tied ropes over the hives. Horace backed the forklift onto the trailer.

I watched Nina walking to her truck. Her bee suit was unzipped to the waist, and the sleeves hung to her knees. She stood on the running board and talked with Horace. She had been at it for seventeen hours now. Horace must have been proud of her, and he must have told her something along that line. Nina stepped down from the truck, and she and Horace stood hugging each other.

On the way back from Mims, Horace talked about his friend Donald Ainsworth, how Ainsworth had a pig that used to come in the house and set its forelegs on the table, how Ainsworth took to bees like a ball of fire. Horace said he didn't have that many friends, and little discipline. He wished he had gone into the Army, because he would have found friends and discipline there.

"You never know when you're not normal," Horace said.

We arrived at Horace's after midnight. When I showed up again the next morning at ten, Horace and Nina were three hours gone, off to a bee yard somewhere.

By April the honeybee tracheal mite had been found in ten states, and was considered to be widespread. The Animal and Plant Health Inspection Service had spent $7 million on the national survey. Since some but not all of the infected colonies had been destroyed, APHIS concluded that there was no longer a basis for the quarantine.

The federal embargo in Florida ended on April 17, 1985, the day after the blueberry bloom began in southern New Jersey. Migratory beekeepers rushed north.

24

Jim Owens is a bee broker in Mays Landing, New Jersey. A broker leases hives from commercial beekeepers and then either makes a honey crop or rents the bees to farmers for pollination. Jim Owens rented hives to blueberry growers in southern New Jersey, thirty-one of them, when I visited him.

I went to Mays Landing in April after the temperature had risen into the nineties and stayed there for a week, quickening the blueberry bloom. Even though the embargo had ended, the migratory beekeepers were wary of going to New Jersey, where the chief bee inspector was threatening to burn entire loads if he found a single positive identification; this inspector had said he intended to keep migratory beekeepers out of New Jersey. For Jim Owens it meant a great difficulty in filling his orders. When Horace Bell backed out of his commitment at the last minute, Andy Card had promised to fill the gap, but he was struggling to do it. During a morning I rode with Jim Owens and his son (he was called

Jimmy, or Little Jim) they removed seventy-eight hives from a group of one thousand in a blueberry field and distributed them to two farms without bees, just to get something in the field. Owens planned to fill the gap with a load from Card scheduled to arrive that afternoon.

Though Owens had a lot of experience in beekeeping, he was a commercial beekeeper for only one month of the year. His main occupation was as a New Jersey park ranger. "I'm not used to these hours," Owens told me. "This runs me right into the ground. Andy and those guys do it all the time, but I have a desk job. I ride around in an air-conditioned car all summer. Physically, I don't work too much. But come bee season, oh. This might be my last year."

"He won't quit," Little Jim said.

"I'm just about ready to go out of business this year because everything is so screwed up," Big Jim said. "If I promise thirty growers they're gonna have bees and the ban doesn't get lifted in time, it's me that's in the heat. There's a risk factor. I could probably be sued if the beekeeper didn't keep his word. I'm the guy that's responsible, so I'd rather have Plan One and Plan Two, and part of Plan Two is that Andy Card was gonna come up with the bees if Horace backed out."

Big Jim was in his late thirties. He was six feet tall and weighed about 240 pounds. His bald head was well-formed and gleaming. On a forearm in blue ink was a bee. Around his neck he wore a gold chain and a gold bee set with two rubies and twenty diamonds. His eyes were a vivid blue. Above an eyebrow was a small divot, made when a truck he was driving in Georgia overturned and, just before he was flung from the cab, a portable television bounced off his forehead.

Big Jim said he didn't smoke and he didn't drink—women were his thing, he said; he just loved women. He managed a nudist camp on the side, and liked to take his girlfriend out there to sunbathe on the weekend. He was also a part-time police officer for the town of Estelle Manor.

At four-thirty we started out, and drove to the back of a thousand-acre blueberry field, with bushes five feet high and shrouded in mist, sand as white as sugar between the long rows, frogs calling in the dawn. Jim Owens puttered along on the forklift like an old gentleman in a riding lawn mower; at least it looked that way to me, compared with what I'd seen Card's and Bell's men do. He loaded up and moved the seventy-eight hives to a five-hundred-acre field in Cologne and scattered pallets along the access roads, talking through the CB radio, sometimes describing the plant life: "That's Blueretta on the right, hundred percent in bloom. Blue Crop on the left, five to ten percent now." We talked about bees to a blueberry farmer who had a fleet of trucks, a helicopter pad, and three Corvettes—his sons'—in a carport. We drove by field after field of highbush blueberries.

We went to the nudist camp and to visit an old girlfriend who'd also worked bees. We went to Owens's office, in a state park, and he went through a file cabinet looking for a photograph of himself standing on a highway, near a truck stop, among a pile of hundreds of overturned hives. Owens called South Carolina several times through the day to find out where Andy Card's bees were. He wasn't even certain the load was on the road.

Jim Owens got into beekeeping in 1961. He was fourteen, doing carpentry for his father, when he saw an old man working a hive across the street from the job. He walked over, watched, then asked if he could buy a hive. He split that hive in two, in another season bought fifteen more, and picked up some small pollination contracts. In 1970 he bought five hundred hives from the New Jersey bee inspector and rented them to blueberry growers.

Then Tom Charnock changed commercial beekeeping in New Jersey, just as he had in Maine. In 1974 Charnock came out of the orange groves into Atlantic County with three hundred hives, on pallets, moved with forklifts. Owens lost three hundred of his seven hundred accounts that year. He drove out to a farm and watched Charnock's hives and waited (Owens, a game warden, was good at the patient stakeout). Charnock had good bees, and Owens understood good bees. When Charnock returned to the barrens to load his hives, Owens struck up a conversation. Charnock offered to show Owens his operation, in return for a little labor.

Charnock had several hundred hives on a cucumber farm on the island of Abaco, and the two men flew over in Charnock's Cessna. Owens was troubled because Charnock kept dozing off; Owens wondered if he was doing it to psych him out. In the cucumber fields they loaded two hundred hives on a truck, by hand, and then spent a day putting them in off-season sites. The roads were narrow, the hives fell off, and the two men kept stopping to piece them back together. Owens was stung hundreds of times.

Charnock hadn't brought water or food. Again, Owens wondered whether Charnock wasn't trying to psych him out. By the end of the day Owens couldn't spit, he was so dry. They reached a sugar plantation and found a water fountain. Owens, his head pounding with a venom headache, drank and drank. He had never drunk so much water. When he finished, Charnock calmly took a sip.

After a winter of reflection on his experiences, Owens became a bee broker. He decided to add $4 to Charnock's price—a dollar to set them out, a dollar to pick them up, and two more for the trouble of doing it. Charnock agreed to deliver the bees.

Owens leased a thousand hives in 1975. In 1978 he took twenty-seven hundred. After the blueberry bloom was over that year, Charnock needed help getting the bees into Maine. They loaded and set out north from Mays Landing. But they had to keep stopping for water, because the truck radiator leaked. Tires blew out, because there was so much honey in the hives.

After the move Charnock asked Owens to return to Maine and work with Charnock's girlfriend, putting together 150 hives that had been torn apart by bears. Then Owens helped with the move to North Dakota. He drove the trailer trucks, he dodged scales, he cajoled his way into gas stations. One of Charnock's other drivers pulled off the road into soft sand, and the hives, tied with ropes, shifted and leaned. The tires on the left side of the truck lifted off the ground. Owens hired a fire engine to wet the bees down. He hired two wreckers to pull from the side, and a wrecker to pull from the front. Then the driver told Owens he'd never driven a semi before and that he wasn't going any farther.

They found another driver and made it as far as Ohio when a head gasket blew. Owens told the other two drivers to go, that he would get the truck repaired. He gave them most of the cash, and kept two checks from blueberry growers.

Owens:

"I got in the truck, it was barely running, and I drove around until I found an old garage. I walked up to the guy and told him I was really in a bind. I told him I'd give him twenty-five bucks if I could unload the truck in his yard for a few days. He said to go ahead. I got nine pallets on the ground when the police drive up. They're beeping the horn at me and running the sirens and lights. I went over and said, 'What's the problem?'

"He said, 'You can't unload those bees here.'

"I said, 'Mister, it's eleven o'clock in the morning. I've pulled the net off those bees. I got them half-unloaded. I cannot put the net back on and leave, because you'll have worse problems than if I unloaded them.'

"He said, 'You're going to take those bees out of here. This is a residential area.'

"I said, 'There's no houses around.'

"He said, 'Up the road a ways they're stinging people.'

"I said, 'They'll calm down. Just let me get them off the truck.'

"He said, 'You either load them back on the truck or you're going to jail.' And with that I said, 'Well you're going to have to put me in jail.' I walked into this whole mess of bees. I figured they weren't going to go in after me. I got into the loader and unloaded that whole tractor-trailer load of bees and I sat in the loader for two hours and they sat in the car for two hours. Gradually it died down, and then they got a call and they left. I went up and down the street giving away honey, and I didn't hear any complaints after that."

The repair cost $500, but Owens's checks were no good in that Ohio town. He had to hitchhike to New Jersey and cash the checks at his own bank. After delivering the bees to North Dakota, Owens went back to Maine for a final load of three hundred hives.

Owens:

"I was coming down the New York Thruway and got to the New Rochelle toll plaza when I blew a hole in the radiator. The gauges weren't working, so I didn't know what was happening. I pulled off to the side of the road, and backed into the toll plaza.

"I made telephone calls. I couldn't get in touch with Tom, I couldn't get in touch with anyone. I called Mays Landing, because there was an old GMC tractor, an old crackerbox Tom had here. It didn't have any windows in it. I tried to get someone to bring that up there but nobody could bring it.

"I went back to the truck to pull the nets, stepped in between two pallets, lost my balance, fell off the truck, and broke both heels. I was in agony, and I didn't know what to do. I had to try to climb up there and get the net off. I couldn't leave it on.

"I hobbled up to the toll plaza. I didn't have enough money to go to the hospital, so I asked when the bus came through. The guy at the toll plaza said no buses would pick me up on the turnpike. I said I was going to hitchhike. He said I wasn't allowed to hitchhike. I told him he was going to have to stop me. A tractor-trailer driver picked me up. My feet were swelling, my shoes felt like they were going to pop off. When I got into Hammondton, my sister came and picked me up. She took me to the medical center and they put casts on my feet.

"I went to bed. But that morning the police started calling the house.

They had seen the name on the beehives, called Florida, talked to Charnock's girlfriend, and she gave them my number. They told me to get the bees out of there. I told them I couldn't do it right now. They said they were going to call in the National Guard to burn them. I told them there were thirty thousand dollars' worth of bees on that load, and if they killed them somebody was going to be responsible. They said I was in trouble.

"Tom had this old crackerbox. The windows had been knocked out of it and it had a dead battery, but we got it started. We got to the toll plaza at ten o'clock that night. There were guys waiting around the truck. I knew they were waiting for me, so we sat back and waited until about two-thirty that morning, when they finally left.

"My friend grabbed the net and threw it on the truck. He drove the Transtar out, we backed the crackerbox under the load, and we took off with the bees. We got to Hammondton at 5 A.M. I pulled down a back street. I just had to get some rest, had to get to bed, but after an hour the police called me up. They were demanding that I move the bees right away. So I had to go down and move them. I unloaded them along a forest firebreak strip."

After his heels mended, Owens went back to North Dakota. Charnock wanted to move his bees across the state to another honeyflow, and he was short on drivers. Two workers had abandoned him. One had left a forklift in a field and stolen Charnock's credit cards. Owens made the cross-state move and then helped move the hives along the third leg of the triangle, back to Florida.

On the second run Owens and another driver were heading south, through Indiana. Owens turned in to a truck stop, and saw that the truck behind him had no hives on it. It was trailing ropes. Beehives were all over the road. Police closed the road and Owens, with the help of some local beekeepers, spent the next three days putting the hives back together. He hired a forklift, loaded up, and headed south.

Owens:

"I went back to North Dakota for another load, and then in Wisconsin a trooper sees me. He made me follow him to the scale. I didn't have fuel permits, and I was overweight for a single axle. It cost me a thousand dollars to get off the scales. They shut the station down and kept me on the scales all that day. I called Charnock, and he said to switch trucks with this other driver who was driving a tandem axle truck back up north.

"I met him in Black River Falls, Wisconsin, four o'clock in the morning, and we switched trucks. They took my truck up to North Dakota for the last load. I got in the truck and took a nap. At daybreak I started to drive.

"Well, this truck hardly had any brakes at all. They ran it out there deadhead, empty, and they didn't need any brakes. Here I had a big, heavy load, and these brakes were really bad. I took my time getting down the road a ways. I'd been to a couple of places and wasn't able to get them fixed. It was the heat of the day so I figured I'd go until night and try to find a brake shop.

"In Tennessee there's a mountain called Mount Eagle. They told me that if I went up there I could get the truck fixed. That's seven miles up, and all the way down the other side there are runaway-truck ramps. You gear way down to get down that mountain with a truck with good brakes. If you do have a problem, you hit these runaway-truck ramps with this real soft sand and that bogs the truck down before it goes over the side of the mountain.

"I got all the way up to the top of this mountain and I got this truck in the shop. They worked on it for three hours, and then they told me they didn't know what was wrong with it. They told me about this place in Marietta, Georgia, that's supposed to be one of the best brake shops in the country for semis. I had to find somebody to hook up to this load. I paid him fifty dollars to tow this load down the mountain for me. I went bobtail down.

"At this point I didn't have hardly any brakes at all, just trailer brakes. When I went to turn off the turnpike in Marietta, I slowed down, but I didn't slow enough to make the comer. I ran up on the curb. When I hit the curb at fifteen miles an hour, the trailer came forward. It knocked me and the chassis off the tractor. I went down flat on the ground. The windshields broke out. My TV fell on top of me and hit me in the head.

"The trailer completely flipped upside down. The beehives in the front end were completely smashed.

"I crawled out. The Georgia Highway Patrol took me down to the station and asked me if I wanted to go to the hospital. I had a cut on my head. I didn't know how many ribs I'd broken. But I told them I wanted to make a phone call. I called Charnock and told him I had flipped this load. He said to get it cleaned up.

"I told the guys at the sheriff's department I was going to try to clean it up. They said, 'No you're not, we're going to get a loader in there and take it to the landfill. You're twelve miles north of Atlanta and that highway is one of the main arteries here. We can't tie up that road.'

"I said, 'There's thirty thousand dollars' worth of bees on the load, and probably another twenty thousand dollars' worth of honey. You have to let me do it, you have to let me get it cleaned up.' I had a case of honey that had come through the accident unbroken, and I gave it to them.

"They put it on TV and on the radio and got some volunteers out there. The wreckers got on each end to lift the trailer and move it to the side. I had to show people how to turn the pallets over. We were working under lights then. And the TV stations got there and they started filming this thing. Everybody was going crazy.

"I'd never been in such pain before, except when I broke my heels. The following day it was raining, luckily. That's the only thing that saved us, because the bees weren't all over everywhere. The police blocked the road off. One guy saw the thing on TV came up from Florida with a loader. We put the pallets back together and took them to a holding yard. It took that whole night and all the next day and night again, but by morning we had everything in the holding yard.

"Then I went to the hospital. They told me I had two broken ribs. I called Charnock's girlfriend, and she came up with a tractor-trailer. I loaded that load up, and they tied them down and netted them and took off for Florida. I hitchhiked over to the Atlanta airport, got on a plane, and that was the last goddamn beehive I saw that year.

"To be a migratory beekeeper, there has to be something wrong with you."

25

We were in a parking lot outside a diner in Hammondton. Young Jimmy Owens and I were standing by the truck. His father was inside, on the CB radio, trying to reach the driver of the semi with Andy Card's bees. The driver had left a message by phone that he had been held up in a traffic jam in Washington. He was close, there had been anonymous comments about a strange-looking load, and Owens was trying to pull him in to us.

"This is my second job since I quit school," Jimmy said. "Until the bees started coming in, I worked at the nudist camp, on the grounds, raking and things. It was fun on the weekends when the girls came out from Atlantic City, but it can be boring."

Jimmy was sixteen, and he'd outgrown the name Little Jim. He was six-four, weighed 250 pounds, had a shadow of a moustache and wavers in his voice, and wanted to go on the road with Andy Card, in September, maybe. This was his first season as a full-time worker. Jimmy did the smoking. He drove the truck in the field while Owens rode behind on the forklift trailer, sending orders to his son over the CB. Jimmy did all the little things that could take so much time—unstrapping loads, removing nets, lowering trailer ramps, smoking, driving. He had a practical, working knowledge of bee behavior and of the landscape.

"This sand is wicked," Jimmy said, after driving too far off the road, as Owens pushed us out with the forklift. "We get stuck a lot. You should see it when it's wet," he said, and went on to tell a story about how when his father was working alone, Owens had gotten stuck in a puddle, and had to steer the truck with ropes while pushing it with the forklift.

"He shouldn't be working out there alone," Jimmy said.

Owens saw another beekeeper sitting at a traffic light in front of the diner, and he called him on the radio. The beekeeper, whose name was Larry, drove up to us. He opened his door, and a wonderful fragrance drifted out. It was the smell of beekeeping, of wax and plant resins, and of bees. Larry had seventy queen bees in cages on the seat of the truck. The queens were from Hawaii—Kona queens, they were called—and Larry had ordered them because he figured they must be mite-free. Larry was operating seven hundred hives, he said, and was going to make increases with the queens, but he soon said he wanted to sell his business.

"I'm thinking about it, too," Owens said. But then they began to talk about making a deal for Owens's loader. Owens wanted a faster forklift.

Then a gleaming, jet-black, raked-back semi arrived, pulling a load of mismatched boxes covered with a sooty-black net. Even though he was only a hundred feet away, Owens communicated with the driver over the CB radio, and following behind, directed him to the holding yard. Andy Card's bees had arrived.

The holding yard was in a sandpit behind a blueberry field. The semi driver stopped, rolled up his windows, climbed into the sleeper with his wife, and wasn't seen again until the trailer bed was empty.

Jim and Jimmy unfastened straps and pulled nets. Bees orbited the truck. Larry, who had followed us to the holding yard, now leaned against his truck and smiled in a satisfied way. "Some of those hives look pretty scraggly," he said. Jimmy explained about Horace Bell's last-minute withdrawal.

"We've got a tough inspector down here," Larry said. "He's said he's gonna burn loads if he finds a single mite. Anyway, if they were to stop these guys from coming in, we'd fill the slack. Within a year we could make enough increases to do it."

"They couldn't do it," was Owens' answer to this. "The bees'd have to build strength on blueberry and the farmers would get ripped off. That's the way it was before. I was there."

It took an hour to unload the 360 hives, and the truck driver wisely moved two hundred yards up the road before getting out to secure chains and put the straps away. Twenty minutes later Owens had 180 hives on his Ford, a monstrous load for such a small truck. (Owens went through at least one axle a year, and kept a spare at home.) They drove to a blueberry field and made periodic stops along the roads, leaving two or three pallets at each site. Jimmy drove the truck and Owens rode in the loader, sitting regally, like a parade master.

We parked next to a pond in a young blueberry field cut out of some pine woods. The temperature had gone well into the eighties, and the bees were very spirited now. When Owens turned a corner on the lift, a whole zigging cloud turned with him. And though it was very hot, Jimmy and I were in the truck with the windows open only a crack.

Owens swirled by. "I never told him this," Jimmy said, looking at his father, "but he's one of the reasons I quit school. He was doing the bees by himself. He has an ulcer, and he was getting sick out there. He almost fell off a tractor-trailer. Last year he called my mom and asked if I could help him, but she said no. She didn't even tell me he called. She didn't want me to miss any school. But this year I decided I was going to do it whether they liked it or not. School started September fourteen. I quit September twenty-nine. In that time I got suspended twice. They didn't like me there."

"So you quit school to do this, to help him?"

"One of the reasons," he said. "Just one of the reasons."

A kind of altruism here? I wondered. And the most unusual kind, perhaps, with the child sacrificing for the parent. But of course Little Jim had the example of Big Jim, and his odd, eccentric stories—Big Jim, something of an altruist himself, as police officer, as game warden chasing armed poachers through the woods (he had stories about this, too), and as beekeeper on the road taking risks for the company. And now, one beekeeper was issuing from the other.

I know it seemed frivolous, a stretch of the imagination, but there was the influence of the hive, too, with its altruism, its natural selection on the family level, with bees sacrificing their lives in the act of stinging, giving off alarm odors to attract other stinging bees; with nurse bees giving up reproductive rights for the benefit of the colony, giving up their own tissues to feed larvae when the incoming food source was short. This was the shepherd following sheep, altruism flowing from nature—but that, perhaps, was taking things too far.

Owens swirled by and swirled by and got the hives stretched out among the fields. The bloom, he hoped, was covered. When it was time to leave, Jimmy climbed into the forklift to cool off in the wind. We drove from the fields, the sun high and bright, the rows of sugar sand incandescent.

"I didn't say much when Jimmy quit school," Owens said. "I mean I did a little, but I could only say so much. His brother is an A student, but that's not Jimmy's thing. Jimmy craves pressure. He has to solve problems.

"You know, you look at him, and he looks like a man. But he's still a kid. He's still sixteen. I have to keep reminding myself of that. I really see it when he eats. And when I get the phone bill—he likes to talk to his girlfriend. But I really see it when he eats."

Owens smiled and let out a sigh. "He wants to be a beekeeper. Maybe it will be like the Marine Corps, like it was when I went on the road with Tom Charnock. He has no idea of how hard this can get. He'll learn if he goes on the road. He needs to know. That's why I want him to get a taste of it."

26

By the second week of May the Massachusetts pollination season was in full gear. Andy Card was bringing hives in from New Jersey and from apple orchards near Billerica, and beginning the move to Maine. Two loads were due in from New Jersey on the afternoon I was at Greenwood Farm; of the thirty-five hundred hives Card sent into New Jersey two thousand had already been brought to Massachusetts. Eight hundred hives were also coming in off apple. Twenty-four hundred would go to Maine, and in a few weeks four thousand would go to cranberry bogs on Cape Cod.

A New Jersey beekeeper had complained about Card's hives and called the inspector, who found a case of American foulbrood. He also took samples for mite tests, and found mites in hives on two pallets. Thus, Merrimack Valley Apiaries joined the ranks of companies with infestations. The New Jersey inspector gassed two pallets, and Card was fined $100. Charnock Apiaries was tested, too. The New Jersey inspector gassed twenty-four offending hives, in the middle of the day. The dead hives were left in place, and the other bees in the area—the depopulation was in a blueberry field of a thousand hives—swarmed into the exterminated colonies to rob the stores (cleptobiosis, this phenomenon is called). There was disagreement over whether the mite was spread further as a result of this depopulation.

But that seemed far behind now. The bloom, the bees, were farther north. Andy Card was in his home territory, and he seemed at ease, despite the intensity of his schedule. "Apple growers are more cooperative," Andy said.

We went out to gather hives, but before we left, I talked to Bruce Betts, one of the part-time drivers who was on his way to New Jersey to meet Jeff Kalmus in the blueberry fields. Betts was an executive computer salesman at the Digital Corporation, and he'd taken a job with MVA, he said, to break the monotony. He liked driving a truck, as it brought him in touch with a different class of people. And although he dickered with Andy on the price for driving, he liked helping him out. "To be honest," Betts said, "I don't need the money."

Betts's face was more smoothly shaven than that of most truck drivers. His plaid shirt, his blue jeans, his visor cap, his boots, all looked like they'd just been issued at the truck drivers' equipment room. Betts was known around MVA as a driver who didn't like to stop. He made straight runs from Florida to Massachusetts. He was also known for his guile. Once, on the New Jersey Turnpike (where migratory beekeepers are not supposed to travel), a toll-collector wanted to know what Betts was carrying.

"Fruit," Betts said.

"What's that buzzing around there?" the man asked.

"Flies. They're attracted to the fruit."

"I met Andy at a party one summer," Bruce Betts said. "We started talking, and he told me what he did, and I kept saying how interesting I thought his business was. Before we got through, he'd invited me to go on a trip to Florida, and I'd asked him to teach me to drive a semi.

"We practiced Sundays, in a supermarket parking lot. I took the test, and much to my surprise passed it. Andy and I were supposed to go to Florida together, but just as it was time to leave, Crystal reached full term with Wesley. Andy had to stay and the bees had to go. I said O.K., I'd do it.

"Andy gave me directions, and everything was fine until I got to the New Jersey Turnpike. Cars started pulling up and beeping and pointing to my load. I had a broken leaf spring. Hives were leaning out. I called repair places on the phone and they'd say fine, come on over, but when I'd get there, they'd see the load and tell me to get lost. I was telling them I was hauling perishables.

"One shop did sell me a leaf spring, though. The temperature got up there, and the bees were beginning to fly, There was a swarm of them around the truck every time I stopped. Then a cop pulled me over for driving a semi in a prohibited area. He made me park in a Burger Chef, leave the bees there, and go to the station with him to pay a fine.

"I called Andy a couple of times, and he kept reassuring me, telling me I'd figure something out. I called a Fruehauf place, and they said they'd fix the spring. I drove right over there. When the Fruehauf manager saw what I was carrying, he said no way. I was beginning to panic. I thought the load was going to cook. The bees were getting hot and I didn't want them all dying on me. So I told the manager that if I couldn't get it fixed, I was going to have to take the nets off to cool the bees. He went inside and found a guy who saw the job as a challenge. I gave him a bee suit, and he set up a lift in the parking lot. He put the leaf spring on right there.

"I was pacing around the yard. The manager noticed that I wasn't getting stung, and he came outside with his camera. He was looking through the lens, and it must have caught the attention of one of the bees. The thing went straight for his nose. He went running back into the shop clutching his face.

"The repair only cost a hundred and fifty bucks, though Andy had to pay another hundred and thirty-five in extermination fees for the bees left behind at Fruehauf."

Betts gave me his business card and left for New Jersey. Andy and I left for the apple orchards.

"I don't know if South Carolina did us any good," Andy said. "This last spring was dry and slowed down the development of the hives. We ran out of pizzazz. We had two thousand cookers, hives with really hot bees. I had to have twenty-five hundred to make that contract with Owens. At about the twenty-two hundred mark I kind of ran out of steam. I had to start shipping packages, and I leased some bees from another beekeeper in Florida. I had to hold some back to take care of this program here with apples. So I got kind of shaky at the thirty-four, thirty-six hundred mark. We wanted ass-kickers in New Jersey, and we sent the best bees there. We sent out bees that were very good on apple, too, but not the class of bees I've sent there in past years."

In Westford, a suburb of Boston a few miles beyond Concord, Andy met Keith Bohne at his farm store. Bohne, an orchardist, had put in a cider mill, and developed a greenhouse.

We rode through a village, turned down the driveway of a Colonial house, and soon came to the crest of a forty-acre orchard. Bohne checked the tree buds ("Viable pollen here," he said, meaning he'd be sorry to see the hives go) while Andy gathered hives. He told Bohne he'd gone too light on pollination, that he should rent one hive per acre rather than one hive per two acres, and when Andy had to drive the forklift, loaded with pallets, over a rocky and rutted road, he told Bohne to place the hives lower the next year. It was only a fifty- or sixty-yard difference in flight, which meant nothing to the bees.

Next we drove to Bolton. "After cranberry," Andy said, "we'll go on loosestrife—joke, joke—and I want to check out the Hudson River Valley, which is supposed to look halfway decent in places. Or maybe around Buffalo we can find goldenrod or something that can support a thousand hives.

I need three or four places. Horace was trying to talk me into going to Ohio. Or western Pennsylvania. This beekeeper by the name of John Thomas said he'd help me find some places out there. I really would like to find a place only a couple of hundred miles from the house. The Hudson River would be fine but there's a bunch of beekeepers out there that hate us. But in New York you don't have to have a mite-free certificate to move in."

We stopped at Doe Orchards, and while Andy gathered the pallets, Mr. Doe came out, leading two collies. It was windy, and apple petals were in the air. Mr. Doe called it petal fall. "They're two weeks ahead of last year," Mr. Doe said.

"What makes them bloom?" I asked, wondering if he'd cracked that code. I hadn't.

"Mother Nature. Flying insects, things that fly around stirring up the air. Last year the full bloom was about the eighteenth, this year it was about ten days earlier. You always figure full bloom is going to be around the full moon. It's one of the ways things function. North, up in New Hampshire in the southern mountains, they're just coming into bloom now. Then you get your hot seasons, 1945, we had full bloom on the fifteenth of April. And they always claimed that the Longfellow poem of Paul Revere's ride in the full blossom of apples in the middle of April was wrong. Could have been. No question about it."

We drove to a hilly orchard in Groton and spend an hour searching for two pallets. Andy complained now and then that apple pollination was in some ways a waste of time, because the orders were small and spread out, and it always took so long to find the hives. But his complaints were lighthearted. It was too nice a day to feel otherwise. The air was cool and smelled of apple blossoms, the view across the hills was magnificent, and the walk was invigorating. I smoked the bees when we found them, and Andy ran the lift. The hill was so steep he skidded, and I had to keep pushing the pallets back on the forks. Two girls walking German shepherds spotted us, and ran.

We drove down I-495 and were on the way back toward Billerica when another beekeeper's truck pulled up beside us. "That's John Thomas!" Andy said. There were hand signals, and at a rest area a few miles ahead they stopped.

Thomas was the man who had invited Andy to Pennsylvania. He was on his way back to Phillips, Maine, and he had eight hours to go. Andy invited him to stay at Billerica, but Thomas said he wanted to get home, that he'd been sleeping in strange beds for fifteen days. He had to kill some bears that had gotten into his hives, too. "Last year I killed a three-hundred-pound bear," Thomas said. "That's not big for Pennsylvania, but it's damn big for Maine."

Thomas talked up the Pennsylvania locale. "Yeah, there's goldenrod August twenty, but some years, and some not. There's a good flow before that, mostly ironweed, Uncle John's wort, tartar flower. It comes. 'Course there's a fellow down there that says if the wind comes from the northeast, there's no flow. Claims the acid rain cuts it down."

They talked about tracheal mites and varroa mites and the killer bees, about chemical treatments they couldn't use, they talked about farm bills and about writing to their congressman. The representatives in Maine, Thomas said, were on the side of the beekeepers. They discussed the federation (the American Beekeeping Federation) and the unwillingness of beekeepers to join—beekeepers are renegades, Thomas said. They talked about trout. Thomas knew of a place where if you didn't catch a hundred six-inch trout in an hour, there was something wrong with you.

"I know where I can catch them this long," Andy said, holding his hands about a foot apart. Thomas responded to that by going to his truck and getting a photo of a 225-pound deer—his deer—and then he got into his truck and continued on to Maine.

"That's a one-man operation," Andy said. "Eight hundred hives. That's a hell of a lot of work. He tried to sell me the whole shebang for a hundred and fifty thousand dollars. But who's going to pay that?" He laughed. "If you had a hundred and fifty thousand dollars to invest, would you buy eight hundred beehives and some beat-up equipment?"

27

When a honeybee forages for pollen, it grasps the blossom's anthers and chews on them, softening the pollen with its mandibles. In actions that are nearly too rapid to see, the bee moves pollen from mandibles and forelegs with a middle pair of legs, and by means of a highly specialized, intricately tooled pair of rear legs, packs the loose pollen into a pellet that may contain a million grains.

On this rear leg is a concave joint. The upper segment of this joint forms the lower part of a press, compacting pollen in the way our elbow joints can press biceps to forearm. The joint is crossed with a bar of stiff hairs that function as a brush, catching pollen swept in from the middle leg. Like a kind of dough machine, the pollen press pushes mashed pollen along the concave surface of the leg, which has, at its rim, a fence made of stiff, curved hairs. The pressed pollen, pushed into the concave basitarsal joint, mounts against this fence. The pollen pushes back not only against the fence but also into the point of a single, long, curved hair called a pollen pin, which runs through the finished load, both stabilizing it and functioning as a spring to keep it close to the leg during flight.

With a pellet on each hind leg, wearing what looks like furry yellow chaps, the bee enters the hive and searches for a pollen-storage cell. Finding one, it drops the back legs over and springs the pellets free. It does no more with the load. A house bee comes along in time, adds saliva, and with its head packs the beebread into the cell.

Though honeybees are incessant groomers, using an array of brushes and a pair of antennae cleaners, and though they constantly rake pollen from their body hair during the harvesting process, they are unable to remove all the pollen from their bodies. Some 10,000 to 25,000 grains always remain.

These are the grains that do the business of pollination. Inside the flower, the stray grain of pollen, adhering to the neck of a stigma, germinates, and forms a slender, threadlike tube that grows down into the ovary, and sends a sperm cell to fuse with an egg. All being right, fruit forms from that fertilized ovary.

Most plants are not self-pollinating—one flower cannot pollinate itself. Apple and blueberry pollen must be transferred not only to another flower, but to another variety of the plant. This is called cross-pollination, which ensures more variety in the species' genetic pool, increasing the chances of adaptation, of quick responses to environmental changes.

A healthy colony of bees, with a population of 50,000 individuals, has approximately 25,000 individuals foraging in the field. Of the 25,000 foragers, 6,000 to 9,000 are gathering pollen. An individual bee averages ten trips a day. In the peak season, when brood rearing is fully under way and pollen is of the utmost necessity, a hive will receive 60,000 to 90,000 loads of pollen (or since each load consists of two pellets, 120,000 to 180,000 pellets) per day.

Depending on the plants, a bee visits 10 to 100 flowers to gather a load of pollen. So if a hive gathers 75,000 loads of pollen with an average of, say, 50 flowers visited per load, then a single hive can reach, in a day, 3,750,000 flowers. But the numbers and effects are far greater because all foragers are transferring pollen on body hairs.

A single pellet of pollen varies in weight, depending on the plant, but a load of maple pollen weighs 22 milligrams. Colony needs require about 80 pounds of pollen in the active season, and a supply for overwintering. If a single pollen load weighs 22 milligrams, and there are 28,000 milligrams in an ounce, then it would take 1,273 loads to make an ounce and 20,363 loads to make a pound. It would take 1,629,090 loads to produce the essential 80 pounds to meet in-season needs. Using the figure of 50 flowers per load, the hive would visit 81,454,500 flowers to gather 80 pounds of pollen.

If the normal flight radius from a hive is, say, two and a half miles, the hive has access to 12,500 acres. In the blueberry fields of Maine (where a grower may expect to gather four thousand pounds of blueberries per acre) honeybees forage close to their hives on their first flights but then range out and select sources that produce more nectar than the tiny blueberry flowers, sources that are, as preferred, higher off the ground than the low-growing blueberry plants.

Some blueberry growers, acting on the advice of blueberry scientists, move the hives every few days to force the bees to reorient themselves, keeping them close to the hives and working the fields rather than the trees.

28

The blueberry barrens of Maine are in Washington County, the easternmost county in the United States. Washington County is in Down East Maine, 310 miles from Boston. It forms the northern end of the Atlantic coastline in the United States, runs along the Bay of Fundy and the Canadian border, and is closer to Saint John, New Brunswick, than to Portland, Maine. Washington County gets about seventy inches of snow a year. Bears don't emerge from hibernation until the end of March, and fruit trees don't bloom until late May or early June. Ninety percent of Washington County is commercial forest, and more than half of the county consists of unnamed townships. Most of the population lives along the coast. There is a large commercial fishing fleet, and fifteen hundred residents own lobster or shellfish permits. There are three large blueberry companies, Jasper Wyman and Sons, Cherryfield Foods, and Northeast Blueberry Company.

Blueberries grow well in Washington County, especially just north of the town called Cherryfield, in an extensive area of glacial outwash plains. The land is flat and hummocky with a sandy, acidic soil, and in these barrens are thousands of acres of dense mats of the lowbush blueberry plant.

Vaccinium angustifolium is a wild plant that spreads along the forest floor or along fields, forming a carpet, seeking open spots of light. It spreads by underground roots, or rhizomes, which send up stems that flower in May and bear fruit in August. The lowbush blueberry is very hardy.

Indians in Maine cultivated blueberries by burning off fields. The blueberry plant, its root system protected from the flames, would flourish after the burn and by the following season produce heavy crops of berries. Maine Indians not only ate them fresh, but also dried them in the sun and used them to cure meat.

In the nineteenth century logging companies laid bare vast tracts of land. In the years after the Civil War, entrepreneurs tried canning and selling the blueberries that grew in these open places, and lowbush blueberry agriculture became a major industry.

In the 1980s, two large packers in Cherryfield were producing ten million pounds of blueberries a year. If you divided up Cherryfield's blueberries among Cherryfield's population of slightly less than a thousand, each person would get about fifty tons of berries. Cherryfield calls itself the blueberry capital of the world.

The blueberry barrens are not farms, in the true sense, because the lowbush blueberry is a forest plant, and the barrens are merely forests trimmed down to this single plant. They are a surprising sight to come upon, especially in wooded and craggy Maine, because they look like tundras, or prairie bogs, or vast wastelands of low-lying tangles of vines.

As you move up close, a fascinating color pattern emerges. The barrens are a patchwork of colonies called clones. The fields are made of brushstrokes of different varieties, which are the basis of cross-pollination and genetic health. Each clone has a different hue—lime green, orange, rust, burgundy, lavender—and each hue is not a single color but a combination of colors. From six feet away a patch may look grayish lavender, but close up the leaves may have a center spade of green inside a serrated, oak-brown border.

When the barrens are in bloom, they have a kind of grain like wood. If you stand and look toward the sun, the flowers are indistinct. But turn and look away from the sun, so that the light is coming from behind you, and the field becomes a carpet of white flowers. There are hundreds of flowers per square foot in a lowbush blueberry patch in the barrens.

Because pesticides are used and fields are still burned, there are few native pollinators in the barrens, so approximately ten thousand colonies of hired pollinators converge there each spring. On the edges of the barrens are many flowering trees and bushes that bloom with the blueberry plants—freeloading, it seems. Perhaps they know the sequence.

The hives are set along the roadsides and at the edges of the fields. Around the hives clouds of bees hang in the wind. Currents whip across the barrens, and bees fight these winds like struggling swimmers. They head into them, they fly obliquely. Because of the temperature and the winds, migratory bee colonies are under substantial stress in the Maine blueberry barrens.

A field of clones, a haze of bloom, a sound of flying bees. The wind can mask this sound, but if you crouch down low, a noise like the crackle of a power plant rises. Even when you can't see the bees except for the lift and soar over the stems, you can hear that electric hum.

There are signs along some of the fields in the barrens that read BEWARE BEES CLOSE WINDOWS AND REMAIN IN VEHICLES WHILE IN AREA. Nonbeekeepers and nongrowers tend to stay away from the barrens and the hovering clouds over the road. But those who know something about bees know the bees are searching, not defending. The signs are meant as much to ward off trespassers and vandals as to warn unsuspecting wanderers. Cherryfield Foods, which puts up the signs, was paying $90,000 to rent four thousand hives in 1985.

Inspecting hives in the Maine blueberry barrens. A bee zooms past the foreground. —GLENN CARD

Many of the migratory beekeepers stay in Millbridge, a few miles south of Cherryfield, in a motel called the Red Barn. When I arrived just before Memorial Day, I saw two beekeepers running down the sidewalk. One had a bouquet of flowers in her hand. She was being carried by the other beekeeper, who was red-faced and laughing. She was Sandy Goddard, and Tom Charnock's girlfriend, but it was all quite innocent, because she saw herself as a guardian to the beekeeper carrying her. He set her down and she shrieked in laughter.

"I had to drive across the country with a sixteen-year-old kid," Sandy said.

"You loved it," said Jimmy Owens.

29

Sandy went into her room in the Red Barn motel and put the lilacs in a bottle.

After tracheal mites were found in Tom Charnock's hives in New Jersey, he was ordered to remove his hives from the state. Charnock went to New Jersey, and being short of help, hired Jimmy Owens to help him load his bees. Then Charnock asked Jimmy if he wanted to join his crew, to become a full-time beekeeper.

"I said to him, I don't care. And then I went to work for him, and here I am. I was supposed to work four days, and I been here four weeks."

His mother told him not to do it, he was too young, but Jimmy Owens remembered what his father had said, that once you work bees it's like a disease. You get in with it and you can't get away from it, you like it so much. He had worked seven days a week, in South Carolina and Alabama. After his first two days off in Maine, Jimmy became edgy and wanted action, so he drove out and found some bear damage. Piecing the hives back together, he got peppered with stings. It invigorated him. He went back to his room and couldn't sleep, he was so hyped.

Charnock had flown Jimmy to a bee yard in South Carolina, and had left him to work with a truck driver from North Dakota and his girlfriend. Jimmy took an immediate dislike to them and decided he was their superior among the bees after they left to go to the bathroom and didn't return for an hour and a half. Bunch of jerks, Jimmy thought—and they stopped every fifteen minutes for soda and cigarettes. Jimmy kept working, diligently arranging and strapping pallets. It was the first of a string of long, exciting days.

They worked in South Carolina, moved to other yards in Alabama, and then readied for the move to Maine. One night they loaded hives on two trucks and early the next day took to the road, the North Dakota driver and his girlfriend in one truck, Jimmy Owens and a hired driver in the other. There were soon problems with the truck Jimmy was in, though, and he stayed up most of that first night to work on it. They drove all day, and again the truck acted up, and again Owens stayed up with the mechanic.

When they got into Maryland, there was a small fire in the truck driven by the couple from North Dakota. They finally put it out, but the other driver, the one with Jimmy, said he'd had enough and disappeared. Jimmy wandered around the truck stop, knocking on doors until he found someone to drive. They moved north through that day and into the night, arriving at Cherryfield at three in the morning. Jimmy unloaded the hives until noon.

He loved it. He was a migratory beekeeper. He would work until Tom Charnock burned him out, he would work until he died. It had been an *experience.* He had driven an array of vehicles. Front-end loaders. He got a crash course in driving a semi, and though he hadn't done any long-distance driving, he had gone about twenty miles once. And when Jimmy and Charnock were flying to South Carolina, Charnock dozed off at the controls. Jimmy nudged him, saying, "Tom?" Charnock told him which gauges to watch and took a nap. Jimmy flew the plane until they hit a storm. They landed in Cape May so that Charnock could go to the bathroom, and they made another stop in Virginia to eat dinner with Charnock's mother.

On one of the trips from Alabama a driver who had missed a connection for a load of watermelons was persuaded to take a load of bees, after promises that the bees would not get out. But they didn't mean every single bee, and there were a few holes in the nets. At the next truck stop the driver was in his truck and wouldn't get out because there were bees circling him. It happened again later on, at another stop. This time the truck driver, knowing that beekeepers use smoke, had lit some small fires, and like a kind of lion tamer was fanning smoke at the bees, trying to coax them back into their hives.

Jimmy was looking forward to Charnock's arrival in the barrens. In New Jersey and South Carolina they had barbecued chops and steaks and drunk beer. In North Dakota they would go camping, and bowling, and to concerts. Jimmy would get his license, first thing. Jimmy said he felt sorry for Tom Charnock, who was having a rough time of it in beekeeping. Jimmy couldn't understand why, as he put it, people were trying to screw Tom.

Jimmy had ridden shotgun with Sandy Goddard on this trip to Maine. Again the truck had problems, something to do with piston rings. From Florida to South Carolina the truck had gone fifty-five miles an hour, but after they got the pallets on Sandy couldn't get over thirty-five, even with a tail wind. She wasn't prepared for the rigors of truck driving, especially on narrow roads, in the rain, with other trucks going by and kicking up thunderous sprays of water. There were tornado warnings.

Jimmy Owens watched Sandy, as he described it, spazzing out and going nuts. She stopped when they went around corners, screamed when trucks passed by, and cried when it was all over. He would make her pull over. He talked calmly to Sandy, which helped. He made her call her mother. That seemed to work best.

Sandy was twenty-seven, blond, beautiful, and gregarious. She had given up a job with an advertising firm to go into beekeeping: "Am I nuts? Huh? I mean, you should see my mother. She flips out. 'You're doing what? You're driving a truck? You're running a Bobcat? What's a Bobcat?' "

She had met Tom Charnock in a bar, where, on a dare, she had picked up a man for the first time in her life. When she asked Charnock to dance, and he got off the barstool, she noticed how short he was and wanted to laugh, but didn't, of course. When Charnock smiled ("all teeth, you know. Heh. Just smiling away") she felt overwhelmed. Sandy walked away after a dance, but Charnock followed her. He wouldn't go away, she remembered. He kind of attached himself to her.

Then, after they'd dated for a while, Sandy came down with a case of viral meningitis. It wasn't diagnosed immediately. She lost strength, became depressed, she told her mother she was going to die. Charnock told her she wasn't going to die, that she was going to get better. He told her she had to think positively. He told her to quit her job, and he took care of her. When Sandy did finally recover, she saw it was Charnock's attitude that had pulled her through, and when she gathered her physical strength back, Charnock gave her a copy of *The Hive and the Honeybee.*

Sandy hadn't known where honey came from. Mosquitoes scared her. When Tom Charnock asked her to come out to a yard and smoke hives while he loaded, and Sandy got out of the car and saw just how many bees were in the air, she jumped back in. Charnock told her she had to overcome her fear of bees. He handed her a bee suit, gave her a smoker, told her what to do. Sandy walked cautiously up to a hive, smoked it, and ran back to the car. Charnock walked over and told her she had to stay out of the car.

And so she learned. Sandy was inspired by Charnock. Most people squashed bugs. Charnock made her step back and say, hey, these are gentle creatures. He showed her the pollen baskets, and he showed her how bees dance. He also showed her how to drive a truck. In North

Dakota Sandy found a role model in Judy Carlson, the state bee inspector, who could tear open most any hive and remain cool.

Eventually Sandy took a solo trip from South Carolina to Texas, with a load of pallets that belonged to another migratory beekeeper. Things went fine until she crossed the Mississippi and one of the bands on the load snapped. Sandy thought, no problem, I have another band. Then the windshield wipers gave out. She had to drive around a small town until she found someone to fix them.

In Lake Charles, Louisiana, there was a steep bridge. Gravity had its way, and the pile of pallets shifted forward over the cab. Some shifted onto the mirror frame, making it impossible to open the door. It was raining, it was one o'clock in the morning, there was no CB radio. Sandy stopped and yelled at passing trucks. She wondered what they were thinking. The police blocked off the bridge and told her they were going to throw the pallets into the water, but Sandy talked them out of it. She drove off the bridge at about five miles per hour, and after a few hours at the police station (they made her wipe the dirt off the door stencil) she continued to Texas.

Sandy, Jimmy, and I were in the Red Barn restaurant when Andy Card arrived.

"Don't forget," Sandy was saying, "Tom is forty-three now. He's been keeping bees for twenty-eight years. For someone to have that enthusiasm after all those years, it's refreshing. It does something for you.

"When you're a beekeeper, you have to know about trucks and how to run them. And you have to know about building and carpentry. And you have to know the weather. And Tom flies, he has to know that. And he has to deal with people. He has to do all the paperwork. He has to know about bees, and has to know about bears, and he has to know about alligators, down in the Everglades."

Jimmy had finished a plate of halibut. He sipped from a bottle of beer.

"What's your pay schedule?" Andy asked.

"You mean how much do I get paid?" Jimmy wasn't sure he should answer that question. "One seventy-five," he said.

"And all the hours he can get out of you," Andy said.

"I don't care."

"You're young."

"On the way up here I didn't sleep for five days."

"I guarantee you," Sandy almost shouted, "if I stick this bee business that within the next five years I'm gonna come up with something, some better way to haul bees than on flatbeds, because it's a crock of goddarn horse hockypucky, the way we haul bees right now."

"Well enlighten me, darling," Andy said. "How do you plan to move them, rail cars, ships, planes? Flying boxcars?"

"There's a better way than these goddarn flatbeds. Loads shifting, blowing in the wind. Flatbeds. Uhh. Horse huckeys."

"Maybe we should use the subways, huh?"

30

In the morning Andy and I drove to Cherryfield, and with the president of Cherryfield Foods and an agricultural engineer, went into the barrens to assess hives. It was a pleasant ride, north from Cherryfield, with the houses spaced farther apart and the trees growing shorter and flatter to the ground, until a huge blueberry field opened up, a field that spread to the distant mountains.

"Blueberries as far as the eye can see," the engineer said. His job, presently, was to track the bloom and direct the movement of the hive trailers. He had worked out a system of overloading a field, putting 240 hives on forty acres, which he said could pollinate a field in four hours.

"The topography here," he said, "on a macroscopic scale, is large," which meant it looked flat but wasn't. When we stopped at the first group of hives, this engineer dressed for invincibility—coveralls, hat, veil, gloves, and several courses of tape around his pant legs.

This was the first year that the management of the blueberry company had gone into the field to inspect hives (previously, they had hired an independent beekeeper to rate hives), so Andy was not only laying open his colonies, he was teaching beekeeping fundamentals.

There were not enough veils to go around, so I cinched up the hood of my windbreaker. It wasn't long before I took a sting on the eyelid. The exchange between pollinator and grower went like this:

"Now what are you calling brood?" management asked. "That there?"

"No, that's honey. The brood is there. Look at that—not a drop of nectar." Andy's implication was that the blueberry flowers were not producing nectar.

"She's spotty there."

"Yeah, it's spotty, but we're just talking about frames of brood. Two, three, four, five, six." This was a minimum requirement, six frames of brood in an eighteen-frame unit.

Andy wanted a show of force. He sprung open a bearded hive, and I cinched my windbreaker opening from grapefruit size to orange.

"That guy's kind of spotty," management observed.

"It's full of fresh eggs, though. You've got to count that as brood if they're full of fresh eggs."

"See, I have a problem seeing the fresh eggs. You're telling me to see something I don't see. Let's see a fresh egg."

Andy held the frame at the proper angle, so that sunlight reached the bottom of the cells. This was always a point of wonder, seeing the tiny egg in the cavernous cell.

From management, the sound of discovery: "Oh."

The education continued, through the morning and well into the afternoon. We traveled through the barrens, by thousands of acres of blueberry fields, by a fifteen-hundred-acre peat bog called the Great Heath. From hilltops we had views of the ocean and the small bays, and Cadillac Mountain, rising above Bar Harbor. We inspected a dozen of Andy Card's sites, and by request, since we were there anyway, a dozen apiaries owned by other beekeepers. One or two were stocked with more

Wesley Card with bee inspectors for Cherryfield Foods, counting frames of brood during field inspection in Maine blueberry barrens. —GLENN CARD

populous colonies than Card's, but a few others, brought in at a cutthroat price by a beekeeper who had missed a crucial orange crop, were far worse. It was an exhausting process, finished off with a final blow when the engineer told Andy that six of his hives had been exterminated by the Maine bee inspector. He swallowed his anger at the sight of the dead combs and the pile of boxes. We left for Millbridge.

"These beekeepers are out here gambling," Andy said. "They're all gambling that they're going to make enough honey to make the low rental price palatable. See, it's a honey gamble, that's what this whole thing is about. But the more money you got invested in that hive, the better. That's the secret of this business, how deep you dare to reach in your pocket in the spring.

"Some people think we truck these bees down to Florida, play all winter, bring them up here in the spring, make big bucks, and play some more. They don't know I work three thousand hours a year, and that Jeff and Dale put in three thousand hours. That works out to a seven-day week.

"I hope they saw out there today that there is a bottom to the thing.

Twenty dollars a hive, they thought, was not an unrealistic amount. But twenty dollars gets pure, unadulterated garbage, and that's because it takes a certain amount of dollars to make a hive of bees. If you make a hammer, it costs so much to make it out of steel, but if you make it out of plastic, it might look like the one made out of steel, but you aren't going to drive any nails with it. Bees and feed are interchangeable. You can raise bees by spending dollars on feed or you can buy bees that someone else fed. You hedge on which is more economical.

"At the present rate of inflation we're losing money, and we can't raise prices. We can't even get them to pay the same as they've been paying the last four years. Never mind paying for bear damage. The price should be thirty to thirty-five dollars a hive. That's what it costs to do it." Andy was getting $29.

Andy ate at the Red Barn, and then he left for western Maine to do some trout fishing, after that, back to Billerica, to persuade the bee inspector and other officials to let him pollinate in the cranberry bogs.

There was a new arrival at the Red Barn, an old GMC crackerbox semi, with "Charnock Apiaries" stenciled on the side.

It had been an especially difficult year for Tom Charnock. He had been the first beekeeper from Florida to have a positive finding for tracheal mites, after late-summer inspections in North Dakota. He had come up with resourceful strategies to stay mite-free, gassing off several hundred hives in North Dakota, buying and shaking mite-free bees into the hives, and then moving them to isolated grounds in South Carolina. He also sent empty hives to Alabama and paid a breeder to stock them, setting up a second mite-free outfit. He moved the remainder of his operation to Florida for an orange crop, fully expecting the quarantine to last. In fact, Tom Charnock was one of the few migratory beekeepers in Florida who wanted the quarantine to last, because he didn't want his mite-free operations to get reseeded by infested operations during the northerly run.

Charnock had devised an intricate system of expansion and contraction to eliminate tracheal mites in his own operation. He wanted to reduce hives in Florida by three to one, move equipment to Alabama, increase by three to one, and keep exterminating bees and exchanging equipment until he had come up clean. But he wondered whether all such attempts were futile, because increasing numbers of beekeepers were carrying mites. Charnock knew that it took only one stray bee, drifting from a passing truck, to enter another beekeeper's apiary and gradually infest the whole operation.

Tom Charnock was among the cleverest of beekeepers. He had good sense for both bees and business, and he had made money at it from his teenage years on. He had adventurer's blood, too, using a plane to scout out territory like a Big Daddy Bee. He was the first to run the triangle from Florida to Maine to North Dakota. He was the first in the East to convert from hand-loading to forklifts, after watching Dave Emde, from the Emde beekeeping family, move palletized hives.

Charnock was a beekeeper whom growers, especially in the early years, saw as a godsend, because his hives were so good. He arrived from long distances on time, one trailer truck after another, piled high, bristling with bees. Charnock was a skilled driver. He avoided scales and inspection stations, went for years without being stopped or fined. Charnock never intended to have overweight loads, but it's part of the nature of beekeeping that bees make honey, and hives become heavy, and an underweight load in Florida can be an overweight load in North Dakota. To Tom Charnock it was part of the skill of migratory beekeeping to skirt weigh stations, just as some other professional drivers often did. And it was part of the economic nature of the business, a matter of the bottom line, that fuel permits and other such expenses were sometimes greater than the fines incurred for not having them.

But Charnock's undoing was that although he could delegate responsibility, he could not delegate cleverness, and he could never hire anyone as smart or dedicated as himself. He would give elaborate maps to a driver, showing where to get off a highway to avoid weigh stations, but then the driver would space out and drive right on the scale and Charnock would have to pay a $900 fine. Charnock didn't hire year-round help; his employees were seasonal, picked up here and there, off the street sometimes, a practice that saved money but also had its hazards. This year, his ambitious multistate hivekeeping system would prove impossible to bring off because of the limitations of his labor force: a truck driver working bees for the first time, a sixteen-year-old high school kid, and a girlfriend.

Tom Charnock and I talked for a while in the Red Barn. It was noon. Lilacs were dipping in the breeze outside the window, and Tom sipped the crab stew he always ordered at the Red Barn. Tom had bright blue eyes, a compact and fit frame, and longish hair that made him look a bit like Ringo Starr.

"I didn't get an orange crop this spring," he said. "That's two years in a row, the first time it's been two years. I can't get too big a crop in North Dakota because I'm not going to take them all out there. I'm just going to take the ones that hopefully I've kept mite-free."

He turned to Sandy. "Did you hear the rumor that was going around that I was going to leave them in Maine year-round?"

"I tell you, Thomas, you could write a book on rumors." Sandy looked at me. "He's just a good old boy, a southern gentleman."

Because Charnock had many trappings of success—a Corvette, a Ferrari, a Cessna—beekeepers were suspicious and resentful, and untruths abounded: Charnock the Howard Hughes-style recluse with a Lear jet who never went into a bee yard, Charnock who had just illegally cut up ten acres of forest in the Everglades and made hive bodies from the wood, Charnock who was put in jail for bringing mites into New Jersey. Strings of tall tales.

"I hear them," Tom said. "Some of them are so foolish I don't even listen to them."

I wanted to ask about business, about beekeeping economics. "What are your average monthly operating costs?" I asked. But my question set something off.

"Nothing's average. Who invented the word *average*, anyway? If we were all average everybody would be alike."

Charnock grew up in Eastville, Virginia. In 1957 he noticed a half-built beehive in a high school shop class, and the components, the smell of the wax, fascinated him. Charnock asked the instructor if he could build a hive of his own. The instructor obliged, and told Tom where he could catch a swarm of bees—there was a colony that lived in a birdhouse near the school and cast a swarm every year. They were descendants of Dutch heather bees, still the dominant strain around Cape Charles, Virginia.

The next year, when he was sixteen, Charnock bought five packages, increasing his hive count to six. He met a cucumber farmer who had a hive but could not take care of it, and made a deal whereby the farmer would pay him $15 a year to manage the hive. Charnock would get the honey, and the farmer would get pollination.

He liked what he was doing, and he liked deriving income from his hives. He went to South Carolina and bought one hundred hives, at $10 each. When he gave the farmer a cashier's check for a thousand dollars, the farmer was so grateful he threw in ten more hives. Charnock took the load to Cape Charles, and by renting them to a cucumber grower for $20 per hive, doubled his money. The wonderful thing about it, he thought, was that he still had the hives.

By the time Charnock finished college he had six hundred hives. His wife was still in school, so while Charnock waited for her to finish, he not only ran the beekeeping business, he leased a Texaco station and opened a used-car lot. In 1968 Tom liquidated his auto business and moved to Florida with his wife and his colonies and a thousand pounds of honey in five-gallon cans. He sold the honey for 10 cents a pound.

The following spring Charnock became a migratory beekeeper. He ran up to Virginia and pollinated for the cucumber growers, carrying 188 hives per trip on a flatbed truck. He unloaded the hives alone, by hand, and alone again in mallaluca swamps in southern Florida, under extreme heat, and got so dehydrated he went to an A&W Root Beer outlet and drank three giant mugs of root beer straight down. And then Charnock saw one of the Emde brothers, down from North Dakota, running a Bobcat.

He looked for new territory. Through an acquaintance in Florida, Charnock met a blueberry grower from Maine who was in West Palm Beach to visit his aunt. Charnock paid the grower a visit, and gave him a sales talk on pollination. The grower had never rented bees, but Charnock made a deal—250 hives at $12 each. He bought a Bobcat, a ten-wheel truck, and in 1972 went to the blueberry barrens. Andy Card's father was there, Charnock remembered, loading little old hives by hand, chained loads of bees. Already it looked so primitive. The next year he brought five hundred hives. Within a decade he reached a peak of four thousand.

After his first trip to Maine, Charnock found out that a blueberry grower in New Jersey was looking for bees. The bees there, Charnock heard, weren't any good. It was the first of May, and the bloom was already under way, but Charnock put a load of three hundred hives together. He went without permits, and he drove through the woods along logging trails to cross the Florida state line and bypass inspection stations. He had the bees in the fields by the fifth of May. New Jersey blueberry growers began singing the praises of the man with the Florida bees.

In 1973, and for the following three years, Charnock migrated to South Dakota. He tried Minnesota for a year, but the beekeepers there, most of whom kept twelve to eighteen hives in a single yard, and had never seen five hundred hives in a pasture, accused Charnock of trying to take over the state nectar sources.

Charnock tried a year in Ohio. He filed for the necessary permits, and placed four thousand hives in a canyon along the Lake Erie shore, a snow belt and good goldenrod country. Again, the local beekeepers banded against him, and he was accused of trying to steal their honey crop.

We were on our way to the airport in Bar Harbor, and Tom was talking about the early reactions to his odd, and in those days, uncovered, cargoes. "I can't believe that I hauled these bees up and down this country for several years with no nets. Every year you'd have to pick different places to stop. It didn't bother me, stringing bees out all over the country. I just wouldn't go back there anymore.

"One time I had a ten-wheeler with a tag axle—an axle in the rear that doesn't pull. After I finished with the bees here in Maine, I headed to Cape Charles, Virginia, to put them on cucumbers. I was on Route 113 coming through Delaware, trying to get around the scales, and this tag axle broke. There was this building, and this nice big parking lot, and I was right beside it. I unloaded the back half of the truck, just filled that parking lot with bees. I set the axle on the truck and went down to Cape Charles with the front half of the load still on the truck. I came back a few days later to pick them up and come to find out I had left those bees at a dance hall. I more or less disrupted the Saturday night parties.

"Then the cops on the Maine turnpike pulled me over one time, back in the early seventies. I was leaving Maine with a semi-load. I stopped at a plaza, went in and got a hamburger, quick. Left about five million bees, I guess. The cops caught me about a hundred miles down the Maine turnpike. They said nobody can get to their cars up at this plaza, everybody's stuck in the restaurant. They said, is there any way you can go back and pick the bees up? I said they'd settle down after a while and all hang together in a little bunch. I said maybe they could call a local beekeeper to go catch them when they cluster. So they let me go.

"But it's more or less settled in now. Everybody's doing the same thing. There's no challenge in it anymore. Time to do something different. Go to Africa, maybe."

I stopped at the airport gate. I asked Tom if I might be able to catch a ride to North Dakota on one of his trucks. He looked at me—sized me up, it seemed and then smiled.

"Sure," he said. "Maybe we'll teach you to drive a semi."

31

The Red Barn was near closing, and a lone, grizzled man in a green uniform sat at the counter. He was a grower who kept a few acres of blueberries, and he rented bees to pollinate them, but farming was a supplementary occupation. This man, whose name was Phil White, and who wore the insignia of the Maine Inland Fisheries and Wildlife Department, was a game warden.

White dealt with nuisance animals. He tended to problems from nuisance skunks, and nuisance raccoons, and this time of the year, nuisance bears, which he did not enjoy.

There are many bears in Washington County. The bears go into hibernation in late November and when they emerge in March, White said these bears are lean and they are hungry and they look for something to eat. They forage in the woods and rivers and bogs. Some of the bears stray into town. They tip over garbage cans, rummage in garages, pilfer in cellarways.

Then, in the middle of May, ten thousand beehives arrive in Washington County. Irresistible odors drift across the barrens and through the woods.

"Those bees are there and the bears eat them," Phil White said. "Male bears on the roam hit one hive here, one hive there, and move along. But female bears nursing cubs are in a more restricted area. They aren't roaming, and they can do some damage to a group of beehives. A sow bear might take a hive per night."

Andy Card called these missing hives MIAs. He had lost as many as fifty hives in a single yard in Blue Hill. Beekeepers who worked the Maine barrens spent days, and nights, putting hives back together, salvaging what was left of clawed and chewed parts. In previous years growers had paid for bear damage or erected electric fences, but then a glutted market of pollinators had appeared. If Card insisted on $25 for every hive ruined by bears, another beekeeper would step in and pollinate without such assurances.

The Maine Inland Fisheries and Wildlife Department issued permits to bear trackers, who set leg traps near the apiaries and then shot the captured bears. One tracker was working out of Cherryfield, another out of Gouldsboro. They were paid fees by beekeepers and growers. The trackers sold the skins, and they sold the meat.

Just now, June 1, four bears had been killed, but by the time the beekeepers left the barrens, the number would rise to twenty.

White thought the policy was a misuse of resources. "It's unfortunate for the bears," he said. "The state of Maine is wilderness, home for the black bear. Now you take a beehive and lug it right out in the middle of the Maine woods and set it on the ground, and think nothing's going to happen to it, you just got to be a little naive. The beekeeper brings his hives up here and sets them on the ground and doesn't do anything to protect that beehive at all. Then he gets upset when the bear stoves it all to hell.

"Fish and Wildlife Department policy allows a beekeeper to request a permit to set traps around his beehives before there's a problem. A farmer can't set traps on his property until after his sheep have been killed by dogs or coyotes. I mean, if you were a butcher and you lugged a bone outdoors and set it on the picnic table out back of your store and you went back in, and didn't do nothing to protect that bone, and the dogs came around and lugged it off, that would be your problem. You wouldn't kill every dog in town. And I don't think you should have that attitude with the bears up here."

Growers think the beekeeper should protect himself. Beekeepers think the grower should provide a safe workplace. Wrath replaces reason at the sight of a seven-hundred-pound pallet turned upside down, plastic bands clawed off, boxes shattered, combs chewed out, and a trail of frames leading into the woods. Some cruel practices arise, such as strychnine bait, or sweetened sponges that expand and block the intestines. Shotguns go off in the night.

"Bear are our natural resources," Phil White said. "If you're killing a lot of bear you can deplete that resource quickly."

A storekeeper in Cherryfield said they had it all backwards. "I think they ought to shoot the beekeepers and leave the bears alone."

Jimmy Owens left for the fields before sunrise. He cleaned up bear damage through the morning, and he took many stings. This time the venom didn't leave him charged and exhilarated.

"We had a big old bonfire. Burned up broken frames and supers. Saved two hives out of twelve in one place. There were five other pallets. Out of those, three were damaged. We made up one hive from one pallet. We took a big cluster of bees and just shoved it in there.

"You could see some of the comb, the marks of the bear claws where they raked it, and some of the supers where they hit it, the big claw marks in the side. They really beat up on some of them. I don't know how I got into this mess."

Hives after bear damage. Wesley Card is looking for equipment in the woods. —GLENN CARD

"Time for a career change?"

"No, I don't want to quit. I like it too much. I meant that we put those hours in, those ten-, fifteen-hour days. Three, four weeks now, and no day off. For a whole hundred and seventy-five dollars a week."

Big Jim would have been proud of the way Little Jim was working through the mess he'd gotten himself into.

I saw Jimmy leave town later in the day, in a semi headed for New Jersey. He looked as if he'd been caught skipping school and was being escorted back to class. But after a week in Mays Landing, Jimmy returned to the barrens, cleaned up more bear damage, and moved to a clover and goldenrod region in New York, sixty miles west of Lake Champlain.

32

It was July. Merrimack Valley Apiaries had gone into the cranberry bogs and come out again. Jeff Kalmus and Dale Thompson were working in a hayfield by a river in Westport, Massachusetts. In the wetlands along the river were spreads of purple loosestrife, which cast a pinkish purple hue to the heads of the dry and dusky green wetland grasses and cattails. Loosestrife grows profusely throughout wetlands west and north of Boston. In the opinion of some people, it is a nuisance plant, because it tends to crowd out the cattails, but to Andy Card loosestrife was a crucial feed plant. His colonies rested near the loosestrife swamps, recuperated, and stored honey before the move south.

There were five hundred hives in the hayfield. Jeff and Dale and Al Carl, the bee inspector, were working along the rows, inspecting for American foulbrood, part of the requirement for the permits needed to move to Florida. Andy had left for the airport, and was on his way to Buffalo to look for locations. Word was there were good stands of goldenrod near Buffalo. Since loosestrife stopped flowing at the end of July, and goldenrod flowed in August and September, Andy thought goldenrod might be the final piece to his yearly puzzle. If he found sites, Dale would drive the bees there in August.

This was a quiet time of the year, this lull after pollination and before southward migration. Jeff and Dale had spent the morning looking for queenless hives (queenlessness is an acute problem for migratory beekeepers—stressed hives tend to kill their queens, as if blaming them for the repeated disruptions to normal hive life). They had dispensed antibiotics, too, tossing handfuls of tetracycline mix on the frames, leaving trails of powder in the air that made the two look like shamans throwing medicine dust.

The three beekeepers worked along under the sun, and for a while I worked with them. Al Carl had on a hat and veil, but he wore a T-shirt, leaving his arms unprotected, and I never saw him get stung. Jeff and Dale were wearing long-sleeved shirts, but their hands were unprotected, and they weren't taking stings, either. Granted, it was a hot summer day, good for flight, and the loosestrife was beckoning, but there were a lot of bees in the air. And while the three beekeepers didn't take stings, I kept getting them—the bees would slam my ankles and sting through my socks or land on my hands. It was almost a magical thing, the way the more practiced beekeepers seemed to take on an aura. Perhaps through the hours the skin became imbued with the scents of smoke, or beeswax, and the bees came to accept the beekeepers.

Before long I left them. I watched the river, and the spread of loosestrife. It's said that purple loosestrife was brought to this country by English colonists, and that seeds also came in imported wool and then washed from the mills into the rivers. And some beekeepers—an early partner of Andy Card, Senior, was one of them—spread the seed along riverbanks, like Johnny Loosestrifes, effecting a kind of pink smoke of July.

But finding locations near these many stands of loosestrife had become tricky. Suburban growth had eliminated fields and blocked access. Andy Card lost locations to development, which was another reason he had flown to Buffalo. He needed outlets.

There was also pressure mounting near his home yard at Greenwood Farm in Billerica. Card's original neighbors had been perennially understanding, but condominium construction was under way along the Concord River. Andy feared that when he had a full colony count at the home yard, and his bees—as they were wont to do—went out for cleansing flights and left droppings all over the new neighbors' car windshields, he would be in big trouble.

But that problem was not nearly so serious as the need to get his bees out of farms—apple, blueberry, cranberry—that were about to apply pesticides, and onto safe, pesticide-free, and ideally, nectar-bearing locations. Sometimes he had to resort to drops.

The location in the hayfield in Westport was not a drop—it was a semidrop. Andy had approached the man haying the property and asked if he could place a few beehives there for a little while. A few beehives was seven hundred (and for Andy they were, truly, a few), and the while turned out to be three weeks (which is, as we know, truly a while). The man who was haying the field didn't own the land. Andy didn't have time for a full public-relations campaign, only a semicampaign. He had thirty-three hundred other hives to attend to.

I walked the length of the field and then crossed over to the railroad tracks. Beyond the tracks was a conservation area, with walking and jogging trails. I was inspecting a honeysuckle hedge when a truck pulled up. The driver worked for the railroad. His tone told me he was not a friend of the Insect World.

"What's this?" he asked. He rolled up the window until it was almost closed.

"Bees."

"They here permanently?"

"Until the end of the week."

"What they here for?"

"They're getting honey from the loosestrife flowers along the river."

"So?"

"Honey. Bees. You know."

"So?" His tone grew more insistent.

"They're beekeepers."

"Hell with this. I'm getting out of here."

"You can work here."

"Hell, no. How come they're not getting stung?"

It was, as I have said, a kind of magic, but I didn't say that. "They know how to do it."

"End of the week?"

"Yeah."

"I'm outta here." He rolled the window up all the way. Rocks sputtered in his wake.

After a few minutes I walked back to the hives and rejoined the inspection. I didn't want any more stings, though, so my touch was tentative and I wasn't as thorough as I could have been.

We reached the end of the row, and extinguished the smokers. There was a count of the number of foulbrood cases found. Jeff had found one. Dale Thompson had found two. I hadn't found any. Al Carl said he had found twenty-four. Jeff and Dale, who already had the opinion that Carl tended toward the excessive, gave each other confirming looks. Dale rolled his eyes.

We walked to the cars and had taken off our veils when a man in a station wagon drove by us and parked fifty feet up the road. With a fixed expression he walked up to us.

"What's going on here?" he demanded to know. It was the landowner.

Dale said he was sorry. "We already moved two hundred hives out, and the rest are going soon," Date said.

"My phone has been ringing off the hook." He nodded in the direction of the conservation area. "People jog on that road," he said, "and take walks along the railroad tracks. Some of them are very upset with this. If anyone gets stung out here, it's me that's liable. I could get sued. I could lose everything I have worked for all my life." He let silence settle, time for his point to sink in. "That's the first thing I thought of when I saw these bees here." He glanced at the hives. "I'm a lawyer," he said.

Sitting on the seat of his car, tying his shoes, Carl looked up and said to the offended landowner, "You won't have to worry about that when the commies take over because you won't own anything."

The lawyer didn't flinch.

Jeff and Dale, who had been down this road before, started a conversation about bees. They explained to the man what migratory beekeeping was, and told him about the nature of the pollination business and the necessity for quick moves, and about how hard it had become to find locations near Billerica.

"I took a load of hives to a place in Newburyport once," Jeff said. "We had been going there every year. This guy came out of his house and pointed a shotgun at me. He told me I wasn't moving any bees in there."

"What did you do?"

"Turned around."

The man's anger began to abate. Dale took a turn. "I took a load by a flea market, and this guy came out with a pistol. He held the pistol at me and told me to get those things out of there. You ought to see how it is getting fuel on the freeways to Florida when you've got a load of beehives on the truck. They wave you through, or they won't turn on the pumps."

The man smiled, kind of, and shook his head. "They told me a few beehives. This is a few beehives? I couldn't have attracted more attention if I'd tied an ape up in the middle of this field."

"Nobody understands bees," Dale said.

33

In 1910 there were 495 beehives in North Dakota. By the time the hive count reached 708 hives in 1920, a new prospect for commercial beekeeping had begun. A beekeeper named F. C. Bennett had produced 360 pounds of comb honey from a single beehive, sold the honey for 50 cents a pound, and written about it in a beekeeping journal.

North Dakota was ideal for commercial beekeeping. Farmers planted clover, hundreds of thousands of acres of it, to improve weakened soils. The plains weather made for copious nectar flows, with winters of deep and durable snows followed by summers of baking days and cold nights. A two-hundred-pound average was possible.

By 1960 two hundred beekeepers were operating thirty thousand hives in North Dakota. By 1970, though the number of beekeepers remained the same, the colony count rose to fifty thousand hives. In the 1970s interstate highways, forklifts and semis, and a federal support price of 65 cents a pound attracted more beekeepers. In 1985 there were some five hundred commercial beekeeping companies in North Dakota, operating three hundred and fifty thousand hives. And those were only the registered hives. The total hive count was probably closer to five hundred thousand.

When I visited Judy Carlson, North Dakota's chief bee inspector, she was just back from her honeymoon and catching up on a backlog of calls. Her office was on the tenth floor of the state capitol building, which is nineteen stories and rises above Bismarck like a finger in the wind. Carlson had spent a day in court, dealing with the case of the Prettymans, a beekeeping family that had filed suit after their hives had been depopulated. Carlson was attending to the mite-sampling operation, and she had given removal notices to twelve companies, all from Florida. She was attending to farmers who were complaining about drops. A farmer with a son allergic to bee stings and who was about to have a wedding reception for two hundred people woke up to find three hundred hives in his pasture. Another farmer found bees by his oil well and couldn't get close enough to get his pump going. A farmer found an apiary in front of the gate to his bull pasture. Some farmers had threatened to burn the hives, but Carlson told them they could be liable for damage suits. And Carlson often didn't know who owned the hives. "Every beekeeper has to have a license," Carlson said, "and they have to get landowner permission, but some people don't get permission. They think, bees are wonderful, I'll give them some honey later."

In her office Judy Carlson laid out maps of North Dakota across a desk. North Dakota has one of the most organized of systems for controlling beekeeping. The entire state is superimposed with a checkerboard of square-mile sections, and in many parts of North Dakota these grids are a reality, in the form of section roads, which border the fields. Many of these squares were controlled by beekeepers, in that they were registered locations. Each location had been taken on a first-come, first-serve basis, and some of the older companies controlled hundreds of them. Beekeeping locations were precious, and well guarded, and since there was also a radius law prohibiting one beekeeper from locating within two miles of another, each location had implications of greater territorial control. But there was a loophole—if a beekeeper made a pollination agreement with a farmer, he was immune from the radius law.

In 1985 some of the more distressed beekeepers weren't registering with Carlson, and they weren't bothering to sign pollination agreements. The North Dakota Apiary Department was sampling for mites ($40,000 had been added to the yearly budget of $240,000 to cover the cost of processing two thousand mite samplings), and the infested operations were given ten-day notices to leave and, in some cases, were escorted from North Dakota by state police.

Judy Carlson ran her finger over the map, from the Red River Valley across the long-grass prairie east of the Missouri River, and over the shortgrass prairie west of the Missouri, touching down on various clusters of red dots inside section squares. Her finger stopped sixty miles north of Bismarck, at a town called Coleharbor, where a group of California beekeepers had settled in for the summer.

"You'll get lots of wolf stories there," Carlson said.

Interstate 83 passed through Coleharbor, by the commercial center—a bar and grocery alongside a restaurant. Both buildings were at the back of an unpaved, dusty lot. A few residential streets led east from the center. Fields of alfalfa, wheat, flax, and sunflowers spread away for miles.

I went into the grocery store and asked a woman at the register if she knew where I could find beekeepers. "There's beekeepers all over town," she said. She went to the window and looked across the parking lot. "Fred's truck is over there. Go on over to the Eighty-three Cafe." She smiled, and gave me a warning. "Watch out for Fred. He might smart you out. He smarted me out, coming in here. I didn't know much about bees then."

There was one man inside the 83 Cafe. He was wearing a visor cap, and gray hair curled up from under it. His face was weathered and lined deeply and haphazardly. His eyes were narrow, hooded, sly, and he could have been fifty, or maybe seventy-five. He was sitting with a cup of coffee and smoking a cigarette, and when I asked him if he was a beekeeper and told him I was following migratory beekeepers, Fred smiled and shook his head. "Somebody should," he said. "Beekeepers are running for their lives. It's gonna be adios time for most of them."

His full name was Fred Tiffany. He owned sixteen hundred hives, spread out in the fields around Coleharbor. Behind the 83 Cafe was a warehouse Fred had built for his beekeeping equipment. It was the second-largest building in Coleharbor, after the old brick schoolhouse, owned by another beekeeper. "Nobody's cutting a fat hog in the ass this year," Fred said.

"Fat hog?"

"You cut a fat hog in the ass, least you get a slice of something. You get a little honey. This year, I don't care what you come up here with. You come up with damn good bees and they're gonna get sprayed and they're not gonna make anything."

Fred was fifty-two. For many years he had run a glass-installation business in Bakersfield, California, and the company had done so well he sold it and retired at forty-two. Fred's brother had fifty hives, and when Fred learned something about bees, he got kind of hooked—and there was the possibility of big bucks.

But you needed more than fifty hives to make big bucks, so Fred invested $95,000 in a thousand-hive operation and ran it with his brother. It was an undertaking that required considerable skills, though, and the Tiffany brothers were not at that level, so after a year they were down to eight hundred hives. Fred's $95,000 was stuck in the operation and he didn't want to lose any more of it, so he took half of the remaining hives, split from his brother, and went down the road.

At first Fred did what nearly all commercial beekeepers in southern California do—he went into orange groves and he pollinated almonds. Fred migrated to Montana in his first migratory summer, but there was a three-mile radius law there, and it was too difficult to find locations. He wintered in Bakersfield, increased his colony count, and went into almonds again, but he didn't like the looks of his bees afterward. The colonies had been exposed to various pesticides, and they were weak. Four hundred hives made no honey crop at all. Although a thousand hives could bring a quick $20,000 in almond pollination fees, Fred began to think that almond pollination wasn't worth risking the health of his colonies.

In 1978, on a tip from another Bakersfield beekeeper, Fred migrated to North Dakota and settled in Coleharbor. There were only two commercial beekeepers in the entire county then. After a good season on clover, alfalfa, and sunflower, Fred decided to winter in Texas rather than California. He set up near the Louisiana line, where it was wet and swampy and flowering. The nectar flow was strong enough for Fred to make three strong colonies from one, colonies that would hit a peak in the North Dakota nectar flow. Fred moved to Coleharbor in mid-May, though sometimes he sent a splinter operation to Bakersfield for the almond run—almonds provided a cash flow, after all.

Fred figured that if he got a hundred barrels of honey—sixty-six thousand pounds—he could "go down the road." Anything above that was profit. Any less and it was adios time. Fred, it appeared, liked the road. He saw his wife once a year—Christmas, he said. Though he had money in the bank he slept in his truck or, in Coleharbor, on a mat on the floor. And though his daughter worked for an airline, and he had the chance at free flights, Fred wouldn't fly because he had to wear a suit. His wife bought him two suits, for Christmas, probably, but Fred looked at them and said—nahh. Maybe he'd fly to Texas next year.

After tracheal mites were discovered on the Rio Grande in 1984, Fred became a pariah in Coleharbor. By then there were six other beekeepers in town, all from Bakersfield. California had adopted strict regulations concerning tracheal mites. Any operation with a positive find would be completely depopulated, and for a time the ruling was that if one infestation showed up in a county in North Dakota, none of the beekeepers from that county could return to the state.

Fred's friendships among the Coleharbor fraternity were strained after the Texas find, even though he was far from the Rio Grande. "Everybody up here got scared, see, because I come out of Texas and they come out of California. They thought I had the mite. Everybody was all up in the air at me. And then Judy Carlson come out, and they sampled, and sampled, and I was mite-free. Which was good."

Fred Tiffany and I got into his truck, sped a mile south, banged a left, coursed over a hill, went by the house Fred had rented for the summer, and stopped a quarter-mile beyond it, by a sunflower field. We got out of the truck and walked through a field to a group of hives. Grasshoppers popped in front of our steps with sounds like snapping twigs. The sunflowers, just to our west, were all facing the sun, necks twisting to follow its course.

"They're just coming in and dropping these bees," Fred said. "This guy put them right in my location. My house is just up the hill."

He pulled a cover off a hive. "They already took the honey supers. See, there's no honey. There's nothing in them." He leaned over and looked at the hive entrance. "There's no flight in them. They're going to move them today. Now, if they start picking up them bees before everything is in the hive tonight, and I'm sure they'll start about four o'clock, them bees will be flying in the air and end up in my hives. They're a quarter-mile away. And I'll have the mite. If they find the mite in my hives, they'll give me ten days to get out of the state. I may have no choice but to kill them. I've got nine yards, four, five hundred hives, setting within a half-mile of this guy."

We left this yard, banged a left by a flax field, banged a left by a sunflower field. Because of the sections and the section roads, driving in North Dakota is an excursion in straight lines and right angles; it's like driving along city blocks, but without the city. Fred calculated: Maybe he would gas his colonies and take the equipment to Texas. Maybe he would buy bees there, a dollar a frame, a dollar a queen, $13 dollars per unit, $20,000 or so to restock.

We stopped at another beekeeper's yard and poked around at the hives. Bees straggled in and out. It was a sad sight for a beekeeper. "If you had to run with bees like that," Fred said, "it would be unbelievable. You couldn't do it, you know. You'd go broke. I think we're going to have to learn to live with the mite, but you can't come up here with bees like this. It would be adios time.

"Let's look at some of my bees," Fred said. We got into Fred's truck. "I forgot to close these damn windows," he said, referring to the grasshoppers that had jumped into the cab and now sat on the seat and clung to the dashboard and to the windshield, watching us with their oil-slick eyes. "A tough animal," Fred said. "A hungry animal. They eat the sunflowers."

Grasshoppers chewed patches through the leaves on the sunflower plants, and they chewed long gouges along the flower heads. And they weren't the only insects foraging there. The flowers were crawling with sunflower weevils, an even greater problem for the growers. In wild sunflowers, the sunflower weevil maintains relatively small populations, but in commercial plantings the populations explode, growing by bounds year after year, until because of accumulating pesticide costs, the crop becomes unprofitable. Sunflower plantings tend to run ahead of weevil populations.

"What it is," Fred said, "there's one spray plane owner that seems to know all these farmers, and he's hustling business. He's selling spray. Whether it pays, they'll never know. But these grasshoppers and these weevils are scaring them."

We drove into a field of tall grasses above a ravine. On the other side a hundred yards away were cornfields and sunflower blocks. A herd of grazing antelope looked up to watch us.

Some of Fred's hives were six feet tall, with six honey supers. "This yard here was knocked on its ass," Fred said. "It got sprayed twice. No direct hits, but over there about a half-mile away they sprayed." In front of each hive was a pile of bees, poisoned foragers that house bees had dragged from the colony and tossed over the edge of the landing board.

Fred knelt down and scooped his hand through a mound. "See that?" he said. "That's their field force. Two weeks ago it was hit. After they were hit, there were no bees in the air at all." Bees were in the air now, however. House bees had reached maturity—three weeks old, they had taken to the field. But Fred didn't expect a honey crop.

We drove past the ravine and along the section line by the cornfield, and then Fred stopped again, by a group of about a hundred hives. "These hives come from California," Fred said. "They were up north of here until last week. No one gave them permission to set there. They just brought them in and set them in here. Now a section line belongs to nobody. You're not supposed to set them on section lines. The farmers can't give you permission for use of the section lines. These guys went and got permission from one landowner and then set off six or eight locations. They came in with thousands of hives."

There were white streamers of paper on the edge of the field, dropped from a spray plane. "Direct hit here," Fred said. "Sprayed right on them. See the dead bees? See them on top of the boxes? That's totally dead. No flight. Two or three bees coming from that one. They must have been nurse bees when they sprayed. It wigs me out how even those lived."

The spray pilot usually let the registered beekeepers in Coleharbor know where and when he would be spraying, but of course there was no fine of communication between the pilot and these squatting beekeepers. "You know, in California they have a spray service. When you're gonna spray, you have to notify the service, and they call the beekeepers.

You can't just go out and spray. The beekeeper has forty-eight hours. And they tell you what kind of spray they're using. Some of them you don't need to run from."

We drove another mile, banged a left, and stopped next to a cornfield. Fred kept fifty hives there, set on bare ground where the corn had been planted and harvested. The ground was strewn with dead bees, and again, there were mounds of bees by the hive entrances. Yet there were bees in the air, and they immediately came at us, buzzing around our faces. The sounds they made had a different tone from the aggressive noises of guards. There was stress in their sound, as if the bees were frantic, making up for lost time. They were also fighting winds. Coleharbor was a windy place, and there had been only a few optimum flying days.

Fred reached down, picked up a handful of dead bees and tossed them in the air. "I tell you," he said. "There's some bees getting killed up here. I can't move all sixteen hundred of my hives all the time. If they start spraying all over up here, I've got to stop coming."

We got into Fred's truck. "You oughta meet Mayfield," Fred said, and we left for his place. Jim Mayfield was also from Bakersfield and had been the first beekeeper to migrate to Coleharbor.

34

Jim Mayfield's house was set in a grove of trees, the only trees in sight. Fred parked around back, we peeked in the honey house, and then we walked up the steps into Mayfield's kitchen, where he and Fred's brother, Jim Tiffany, were sitting at a table drinking beer. Both men were in their mid-forties, tired and a bit dusty from a day in the field.

Fred told them about the unregistered hives, and together they tossed about possibilities for action.

"Maybe we should just burn those Florida hives. A citizen's depopulation."

"Can't do that."

"Gas them off?"

"How about a slash of red paint on all of them? Mark them so he can't take them nowhere else."

"Can't do that, either. If that guy went and got a lawyer and proved you defaced his hives, he could sue you."

"Maybe we should just shoot the guy. I mean, we're gonna die here."

"No, that wouldn't do it, either."

A smile. "Maybe we should all go to Florida together."

Jim Mayfield had received mail from his wife that day, an envelope full of news clippings from papers in the Bakersfield area. All the stories were about the discovery of Africanized honeybees in Lost Hills, California, forty-five miles north of Bakersfield. It was the first discovery of killer bees in the United States.

Mayfield had spread the clippings out over the table. "That's where we come from," Jim Mayfield said. "Those killer bees are right where we live. Kern County."

"They'll be gassing off more hives," Fred said.

The headlines told the story: "It's Official: Killer Bees Found in Kern." "Bee Strategy Debated; Scorched Earth Policy Out; Top Scientists Gather." "Aggressive Bees Rout Entomologist; Swarm of Angry Bees Chase Scientist Away From Nest." "Bee Battalions Mopping Up Killer Bee Invasion." "Africanized Bees Won't Take Over County Just Yet." "Bee Checkers Get Day Off." "U.S. Enters Bee Fight with Own Quarantine." "Experts See No Quick Solution to Killer Bee Problem." "Invader Bees Reproducing." "Killer Bees Discovered in Lost Hills Domestic Hives." "Bee Parasites Considered Worst Danger." "Beehives Removed in Quarantine Zone Without Inspection." "Officials Locate Beehives Taken from Lost Hills."

The story ran all over the country. It was not *the* killer bee story—that would come in 1990, when the natural front of the African bee reached the U.S. border after migrating through Mexico. Or perhaps *the* story would come when the first unfortunate American citizen died from stings taken from this hyperdefensive strain of bees. The California find was not part of the natural front. That colony of Africanized bees and its descendants had emanated, most likely, from a colony—a hitchhiker swarm—that had settled on oil-drilling equipment in South America, and been transported to Los Angeles and from there to Lost Hills. (Or perhaps the parent swarm settled on a ship moving through the Panama Canal—of fifty-one hundred swarms of African bees captured in Panama during a three-year period, three hundred had been found on ships. One that eluded detection had been found, dead, in Buffalo. Another, alive on a sugar container and thus with a food source, was found in England.)

The Lost Hills story began on June 6, 1985, when a machine operator, working in an oil field in Lost Hills, watched a swarm of bees kill a rabbit and then swarm at his cab. The driver had noticed that the bees issued from a hole in the ground—a kit fox den—and he covered it with chunks of asphalt. He didn't want to report what he'd seen, but his wife talked him into it. The driver, knowing he would be swarmed by reporters, asked for, and received, a promise of anonymity.

The bees were killed, and the nest was opened, revealing twenty-four combs built between two timbers. Some of the combs were five feet long. The nest had been occupied for at least a year. The colony had, in fact, swarmed a few weeks before discovery. It had built three queen cells and was about to swarm again.

Samples of the bees were taken, but conclusive evidence that the bees were in fact Africanized bees took six weeks. The various races of honeybees can look pretty much alike even though they behave quite differently. Entomologists can measure combs and make chemical analyses, but the best tests are computerized comparisons of vein patterns in the wings and the size of body parts—multivariate morphometric analysis.

On the day following conclusive identification a special task force of university and government biologists assembled at the Kern County agricultural commissioner's office and then visited the Lost Hills site. One survey crew began mapping apiary locations by helicopter, and another crew surveyed four hundred square miles on the ground. After a week they had found ninety-seven apiaries and ninety-two hundred commercial beehives. Members of the ground crew conducted preliminary Africanized bee tests—approaching a hive, kicking it, and then waiting to see what happened.

The sampling process began. Fifty bees were taken from each colony and sent to the lab. In addition, inspectors were hired to track and destroy all wild colonies within a fifty-mile radius of the Lost Hills find. All beekeepers were prohibited from moving colonies out of the area until the testing was completed, but one beekeeper was caught moving honey supers with bees inside. He said he'd know an African bee if he saw one; nevertheless, his load of supers was quarantined, the bees were killed, and the beekeeper was fined $1,830.

After a week of sampling, a second colony of Africanized bees was found, two miles north of the Lost Hills find. This colony had entered and taken over a hive of European bees. The survey area was extended forty-two miles.

On August 14 a colony of Africanized bees was found in a hollow tree stump, seven miles southwest of the original find. It had been in place for at least a year. The colony was destroyed, and surveys were extended to 172 square miles.

On August 21 a fourth find was confirmed after a call by a beekeeper from Bakersfield. He had captured a swarm on a water tower, hived the bees and placed the colony with twenty-five others near California State College at Bakersfield. The colony produced a lot of honey, but the bees also stung a lot, which had made the beekeeper suspicious. He had cut out queen cells to keep the bees from swarming—standard beekeeping procedure—and the colony, its population unrelieved, expanded to five hive bodies. After the bees were moved to a buckwheat bloom in the Kelso Valley, the colony swarmed, and when the beekeeper returned to Bakersfield, he called the bee inspector. The colony was destroyed, and the quarantine was extended to all areas within two miles of that apiary. A wild swarm survey was conducted in the Kelso Valley.

The fifth find of Africanized bees was a mile east of the Lost Hills find, in a migratory apiary that had moved into the area on July 23.

On September 6 inspectors found a sixth swarm that had moved into a dead hive in a migratory apiary. The colony had a queen and eighty-three hundred workers and had prepared three queen cells in preparation to swarm.

A seventh find came on September 27 at an oil-processing plant in Lost Hills. The colony was in a steel pipe used to support a retaining wall, had been in place for six months, and hadn't attacked anyone, according to a safety officer. Tests revealed that the bees had been diluted genetically, through interbreeding.

The eighth find was on October 2, five miles east of the fox den. These bees were also diluted, and not highly defensive.

The ninth Africanized colony was found in a tree near Bakersfield College.

The tenth find was in an apiary in the quarantine zone.

The eleventh colony was found in a discharge pipe, and it remained placid while being scraped into a plastic bag.

A twelfth colony was found in a managed apiary and on November 8 was destroyed. And on December 2, after sampling twenty-two thousand colonies, the Africanized bee project closed down, declaring that Africanized bee colonies in California had been eradicated. The first invasion of killer bees was over.

You could say—and some did say—that the Lost Hills find was merely the first publicized find, the first Killer Bee Event, and that the Africanized bee had actually been around for a while already. Colonies with extremely defensive traits had shown up in the apiaries of many beekeepers—some of the Coleharbor beekeepers said that they had been working killer bees for years. Beekeepers would never report an "incident"; knowing the public reaction, they would have been crazy to. They would have requeened—standard practice for changing the temperament of a colony—or, suckers for the honey crop, they would have managed the colony, if it was a super honey producer.

But the Lost Hills incident was ahead of the frontal migration of the Africanized honeybee. At the same time Fred Tiffany, Jim Tiffany, and Jim Mayfield reckoned with mite infestations and news clippings from home, *abejas furiosas* was moving through Central America. Nicaraguan beekeepers had been noticing swarms of dark bees that flew very fast, at a higher altitude than normal. And something else was odd. These swarms flew during the dry season. The Nicaraguan beekeepers hadn't seen that before. Some beekeepers were destroying these colonies of *abejas furiosas,* of *jicotes extranjeros,* the foreign bees, to keep them from mixing with their own.

35

The Africanization of the honeybee populations of the Americas began in Brazil in 1957, when a well-intentioned beekeeper removed the fences from the entrances of thirty-five beehives in an experimental apiary in a eucalyptus forest. The fences, meant to exclude the queens from any group of bees attempting to swarm from the hives, were jammed with pollen, and the beekeeper only wanted to free things up. But the hives contained queens recently imported from Pretoria by a Brazilian biologist who was a member of a project to find a better bee for the Brazilian climate. European bees, their rhythms locked into the temperate zone, were not making big honey crops there.

After ten days, when the removal of the fences was discovered, twenty-six swarms had departed, and in a genetic ascent that has left biologists mystified, those twenty-six swarms begot ten million colonies that, through various other adaptive strategies, displaced most of the European bees in South America. Ironically, the experiment worked.

Thirty years after the initial collapse of the Brazilian beekeeping industry and the subsequent partnership between Brazilian beekeepers and their Africanized stock, honey crops reach five hundred pounds per hive.

The region of origin of *Apis mellifera* is thought to have been in Africa or perhaps in Afghanistan or India. The races of *Apis mellifera* (*Apis mellifera ligustica, Apis mellifera carniola, Apis mellifera caucasica,* and so on) that migrated to the northern temperate zone live in harmony with the cycle of spring, summer, and winter—an extended flowering period followed by a dearth of nectar and a cold period. European honeybees store large quantities of honey, cut off brood rearing in the fall, contract into an organic furnace, and then when spring comes, build large colonies and sometimes swarm. Swarming behavior varies from race to race, with the more northern races, those with delayed springs, tending to swarm quickly so as to be in place for the prime summer flow.

The equatorial, African honeybee (*Apis mellifera adansonii, Apis mellifera scutella,* and so on) adapted to the region between the Sahara and the Kalihari, where nectar flows are triggered not by spring but by tropical rains. The African honeybee responds to the tropical rains with explosive population rearing, so as to be ready for the flowering that soon follows. An African queen bee can lay four thousand eggs a day, about twice the number of her European cousin. During the nectar flow an African bee lives a short life, only twelve to eighteen days (rather than the thirty-two to thirty-five days of the European bee), which makes for a quicker response time. An African bee can be foraging six days after hatching, and with a development time from egg to adult of only eighteen days, the African colony can have a field force only twenty-four days after a rain. A European bee would be only three days out of the cell by then, and eighteen days from foraging age.

One of the surprises of the Africanized bee in South America is that it has shown the ability to do something bee scientists long predicted it could not do—form insulating clusters and live in cold climates, which it has done in the Andes. Maybe the Africanized hybrid acquired the information to winter genetically from breeding with European bees; or maybe the progenitor of the African bee adapted to tropical regions after it had acquired the ability to form thermoregulated clusters in more northern regions.

One obvious but fundamental fact about the honeybee: Without its stinger *Apis mellifera* would have been extinct long before *Homo sapiens* came along and discovered bee space. The phenomenon of mass stinging has kept the honey flowing and the waxen architectures rising for tens of millions of years. Honeybees are preyed upon by mites, ticks, ants, dragonflies, wasps, moths, frogs, bears, and other honeybees (cleptobiosis). In Africa there are honey badgers, and a bird, the honey guide, that incessantly chatters at other honey eaters—badgers, baboons, men. The honey guide leads its prospects to a bee colony, waiting nearby for the hive to be opened so that it may snatch a stray bit of comb.

Honeybees are unique livestock in that, even though they are "kept," a colony becomes feral, or wild, as soon as it leaves the man-made hive and takes to a hollow tree. Nevertheless, most of the European races are products of centuries of selective breeding for combined traits of gentleness and productivity—the excessively defensive colonies were, and still are, killed off.

The process of selection in Africa, however, favored the most defensive colonies, those highly responsive to changes in odor, sound, and movement. African bees are easily disturbed, sting in swarms, and once they have begun stinging and releasing alarm odors, tend to pursue the intruder for hundreds of yards. The European colony tends to calm down in three to five minutes after a disturbance, but the recovery time for an African or Africanized colony can be thirty minutes. There have been many tests—a leather ball takes three stings in a minute from a European colony and one hundred from an African hive—but perhaps the best generality about African bee stinging is that it can be at least ten times greater.

The African bee became the killer bee, and later, in a measure of improved public relations, increased understanding, greater scientific accuracy (and, you could say, melliphonia), the Africanized bee when, moving in radial waves of two hundred to three hundred miles a year from its epicenter in a Brazilian eucalyptus forest near São Paulo, it attacked and killed livestock and humans—all in defense of the nest.

By 1965, seven years after the bees escaped from the test apiary, 70 percent of the European colonies in Brazil had become transformed into Africanized colonies. This process of Africanization was a result of several unique traits of the African bee.

All honeybees swarm, by which one-half to two-thirds of the bees depart with the queen and establish a new nest, leaving behind a virgin queen. European bees tend to swarm once, perhaps twice, in a season, and because of the nature of the summer flow, the new colony must soundly establish itself and prepare for winter if it's to survive.

A colony of African bees, anticipating tropical rains, tends to swarm in multiples, casting one small colony after another, each waiting to bloom after the triggering mechanism of rain. A single colony of African bees can cast dozens in a season.

European bees tend to have an attachment to their nests. When the nectar flow stops, they rely on their stores, which can consist of hundreds of pounds of honey. If a European colony does run out of honey, it remains in the nest, and if the flow doesn't come, it starves there.

African bees respond differently to hunger, what biologists call nutritional stress. An Africanized colony will readily abandon, or abscond, the nest, and will fly for long distances, looking for a spot where it has rained and there is food. In Africa these bees tend to travel along migratory paths, and African honey hunters, long aware of these lines, set out trap hives for the swarms.

A Kenyan scientist by the name of Isaac Kigatiira had people follow these trek swarms on foot, bicycle, horseback, and motorcycle. They lost sight of many swarms, of course, but they tracked enough of them to observe that a swarm stops after about twenty miles, contracting into a cluster. The colony then searches for a nectar source, and if it finds one, it forages for a few days. Then, depending on the efficiency of the forage, the swarm moves on. A swarm tends to travel about sixty miles, though Kigatiira's trackers observed some that traveled as far as eighty miles. After this migration the colony nests, and in time casts off its own swarms.

African bees also practice another kind of swarming, a nest parasitism that has been called commando swarming. About two hundred bees with a queen, a unit the size of a teacup, land on the side of a hive. After dark a group of workers enters the hive, finds the resident queen, and then kills her. With the death of the queen the pheromone conditions alter and the colony gradually changes its disposition to that of queenlessness. Timing is important. A colony will not accept a new queen until they have accepted their own queenlessness; the commando swarm, with instinctive intelligence, knows when to enter the colony with its own queen. Soon she begins to lay eggs, and within a few weeks the colony's population changes, over to the commando queen's progeny, from, say, European bees to African bees, and the colony thus becomes Africanized.

Another process of Africanization is something called drone saturation. Colonies of African bees produce high proportions of drones, compared with European colonies. African drones tend to stay in the air longer and to fly faster than European drones. The result is that in an apiary of 50 percent European colonies and 50 percent African colonies, 90 percent of the matings are with African drones. It is suspected that African drones also indulge in male parasitism—following European drones back to their hives and becoming members of the European colonies, since drones are usually accepted into any colony.

European honeybees tend to be very selective about nesting sites, and there is often extended communication when selecting one. But Africanzed honeybees, though they prefer large cavities, will nest in places European bees would pass up—gutter pipes, street lamps, animal burrows, electrical outlet boxes.

One of the peculiar enigmas of the process of Africanization is that a small gene pool—twenty-six colonies—has resulted in a continent of bees with the trait of extreme defensiveness, a trait that hasn't diminished (though many have claimed it would) as *Apis mellifera brasilia* moves north.

The Africanized bee moved into Argentina at a relatively slow rate, because of the cold climate, but colonies have been found in the Andes above two thousand meters, where the temperature goes below freezing during six months of the year.

Africanized swarms entered Venezuela in 1976, and within one year the bee population had become Africanized. Of the eighteen commercial beekeepers operating in Venezuela, all but two abandoned the business, and Venezuelan honey production dropped from 580 metric tons a year to less than 100 tons. Most of the part-time and hobbyist beekeepers, those people with a hive or two in the backyard, killed their bees or abandoned their equipment after encountering the defensive proclivities of Africanized bees.

Trek swarms landed on a ship in a Venezuelan port and traveled to Trinidad, and Trinidad became Africanized. One Trinidad man, cutting grass with a machete, overturned a piece of metal covering a colony, and was stung to death when, rather than running away, he ran in circles and swatted at the bees.

Trek swarms moved through Colombia, flying over water and moist, low-lying areas. In 1980 or 1981, swarms crossed the Panama Canal. Many swarms settled on passing ships. In 1983 swarms moved into Costa Rica, and the beekeeping industry went bankrupt. The United States Animal and Plant Health Inspection Service spent $1 million to contain the migration in Costa Rica, but was unsuccessful.

In April 1984, residents in southern Nicaragua noticed swarms of bees moving by. By October, 75 percent of the colonies there had become Africanized. Eight months later, 250 kilometers north, and during the dry season, no less, Nicaraguan beekeepers were finding swarms of *jicotes extranjeros.*

Africanized bees migrated into Guatemala and into Belize, and Peace Corps beekeeping specialists began to work with local beekeepers, helping them adjust. In the summer of 1986 a biology student from the University of Miami, working in a Costa Rican rain forest, was attacked and killed by Africanized bees.

By the fall of 1986 the Africanized bee had moved into Mexico. The first confirmed reports came in December. By February there were at least fourteen confirmed finds. Mexican, United States, and Canadian officials met in New Orleans in early 1987, to discuss the dilemma, and a figure of seventy finds was announced.

Trek swarms moved north along both Mexican coasts. Though it was expected that a large population of European bees would interbreed with, alter, and slow the migration of Africanized bees, that seemed unlikely, since 40 percent of Mexico's hives were in the Yucatán peninsula and most of the rest were in the central highlands. There were perhaps fifty thousand colonies in the coastal regions. The Mexican beekeeping industry—Mexico was the world's largest exporter of honey at the time—was in disarray, and the Mexican crop began to diminish. Many part-time beekeepers abandoned beekeeping. Only the largest companies were expected to survive the transition.

By 1990 at least six people had died in Mexico from Africanized bee stingings. The Mexican state of Chiapas was 70 percent Africanized. Quintana Roo was 50 percent Africanized, the Yucatán, 40 percent.

36

Just as the tracheal mite crisis was emerging, Reggie Wilbanks left Georgia and flew to Venezuela to observe Africanized bees. Wilbanks was among sixty commercial beekeepers who visited the Department of Agriculture's field station, run by Dr. Thomas Rinderer. Rinderer and his researchers were studying drone saturation, in an attempt to alter the behavior of the Africanized honeybee.

When these beekeepers, all seasoned professionals who had often taken dozens of stings in a day's work, drove to the test apiary, they were shocked at the response of the Africanized bees. Wilbanks, for one, stiffened with fear. Unable to move, he stood with his arms straight out. The bees covered his camera lens and smeared it with venom.

"Have you ever seen bees swarm?" Reggie said to me. "Piling out, running over each other, running to get out of the hive? Well, down there, it started at the closest hive, and then there was a chain reaction.

They plied out all down the line, but instead of going in the air to swarm, they go at you. One man who'd been keeping bees all his life said he wasn't afraid and was going to go without gloves, but once we got there he put them on right away."

Wilbanks had thought that the Africanized bees were a media hype, and that scientists were getting grants to study them as an excuse to go to South America for work in the sun. But Reggie had changed his mind. The Africanized bees, he realized, would be a serious problem—if they moved into the United States with the traits of the bees he observed in Venezuela.

"We don't have areas isolated enough for the Africanized bees. I thought all those reports of stinging deaths were exaggerated. But after seeing those bees, I can believe it.

"They are going to put us completely out of business," Wilbanks said, "unless we dilute the characteristics of those bees. In South America they encountered little European stock, but in the United States, with a large European stock, they could be diluted. We wouldn't have to go through what the people went through down there. The USDA should get prepared, and not get caught with its pants down like it did with the mite. And the beekeeping industry should be dealing with the public. This could wipe out the industry.

"I'm not saying I'll do it, but better judgment would be to gas the bees and get out of business."

At about the same time Wilbanks went to Venezuela the USDA published a report estimating the losses that could occur if the Africanized honey-bee Africanizes the United States. Written by Robert McDowell, the report was titled "The Africanized Honey Bee in the United States: What Will Happen to the U.S. Beekeeping Industry?"

If the USDA did not have its pants down, it certainly did have its hands tied. Employees did not want to create a panic, so the language of the McDowell report was very cautious. The first paragraph of the report addressed concerns about reactions to the Africanized honeybee:

"The general public has a special fear of stinging insects and a particularly morbid fascination with hordes of stinging insects. A number of sensational news stories and motion pictures about the Africanized honey bee (AHB), also known as the Brazilian bee and the 'killer bee' (a misnomer), have played on this fear. The result has been a mixing of fact, fiction, myth, and misinformation to the detriment of a wider, clearer understanding of the issue."

The paper laid out four scenarios, devised with two questions in mind: How far north will the Africanized bee migrate? and Will it retain its excessive stinging behavior?

The first scenario was the most liberal, making forecasts on the assumptions that the Africanized bee would inhabit the eleven states with at least 240 frost-free days per year—Virginia to California—and that it would retain its stinging traits. The total economic losses in Scenario I would be $53.8 million to $58.5 million and broke down as follows:

1) All package and queen bee producing states would be quarantined. A business producing $16 million a year would suffer losses of $14.4 million, or 90 percent.

2) Of hobbyists, 50 to 80 percent would quit beekeeping. Commercial beekeepers would make 30 percent less honey, for a loss of $22 million to $26 million.

3) Migratory beekeeping would cease to exist, and the cost of overwintering in the north would be $13.3 million.

4) Beekeepers who normally pollinated in almond orchards in California would lose fees of about $1.4 million.

In Scenario II, based on habitation of the same eleven states, but with slightly tamer bees, fewer colonies would go out of production, honey crops and wax yields would decrease less, and economic losses would be $49 million to $52 million.

In Scenario III, based on a territory below a latitude of 32 degrees—basically the states on the Gulf of Mexico—and on the assumption that the bees would remain excessively defensive, 50 percent of the noncommercial colonies would be abandoned, yield among commercial beekeepers would drop 30 percent, and losses would be $28 million to $31 million.

And in Scenario IV, based on the Gulf states territory but with diminished stinging, losses would amount to $25 million to $27 million.

Appendixes in the McDowell report implied greater losses. One section discussed the likelihood that commercial beekeepers could not sustain 30 percent losses in honey crops, that they would not be able to cover operating costs, and that honey would have to go up to 76 cents a pound wholesale to make up the difference. The implication was that many commercial operations would fail and that losses would be much greater.

The report was tentative on the subject of crop pollination, factoring only California almond pollination into the scenarios. This was partly due to data constraints. But an appendix suggested that a lack of pollinators could result in far greater economic losses. A mere 1 percent decline in the harvest of fruits, nuts, and vegetables that depend on honeybee pollination, with a combined value of $3.5 billion, would amount to a loss of $87 million.

The scenarios in the McDowell report carried the assumption that migratory beekeepers could overwinter in northern states by using supplemental feed and supplemental pollen. But that would cost money. And there was also the prospect that honeybee mites would result in increased winter losses.

One section (called Rationale for Assumptions) spoke of some devastating implications, at least for the beekeepers. Beekeepers could adjust to the Africanized bee, it said, learn how to work with it, altering the traits of their stock as problems arose. The more serious problem would be the reaction of the general public (which tended to be misinformed about beekeeping), resulting in zoning ordinances prohibiting beekeeping, restricted apiary sites, and lawsuits from stinging incidents.

And there was a paragraph suggesting the difficulty that people like Reggie Wilbanks might have, breeding queens in the wild, should the air become saturated with Africanized drones.

In 1986 there were attempts to establish a control zone in Mexico before Africanized bees moved in from Guatemala. Senator Charles Mathias of Maryland proposed that $1 million of Department of Agriculture funds be allocated to begin the project.

Dr. Thomas Rinderer, the USDA specialist on the Africanized bee, and his staff designed a plan to counter the processes of Africanization. Their control zone would cover the width of the Isthmus of Tehuantepec, where Mexico is less than two hundred miles wide, and be one hundred miles deep. The plan included bait stations, net sweeps, and feeder hives. The Mexican government would set up highway checkpoints to inspect the trucks of migratory beekeepers, and also educate Mexican beekeepers on Africanized bees.

Workers would trap trek swarms in cardboard boxes, and since the zone was a hundred miles deep, the chances were good of trapping trek swarms during feeding forays. The boxes would cost 7 to 10 cents each. Bounty hunters would be employed to bring in swarms. The zone would be flooded with drones from European stock. There would be a program to control the queen stock in Mexican apiaries.

The project became known as the Bee Barrier, a misnomer, since it implied that the Africanized bee would be stopped at the isthmus. The intention was in fact to slow it down and possibly alter its behavior through interbreeding.

There was disagreement in the American scientific community about the Africanized bee. One theory held that it would bring disaster; another theory said it would be unable to cross arid regions of Mexico;

still another said it would be lost in the vast genetic pool of Mexican bees. Congressmen, uneasy about spending money, especially on something as suspect as a Bee Barrier (*Time* magazine published a cartoon of a bee nuzzling a brick wall), were, as one USDA official put it, "more willing to listen to the bearers of good news than the bearers of doom."

Rinderer thought that the zone would be funded, and that the government would do something about the Africanized bee as it moved through Mexico. "If we don't," he said, "in the areas of the U.S. that become Africanized we can expect hobbyist beekeepers to become very rare. I think migratory beekeeping would be very hard hit."

Funds for the Bee Barrier were approved in the Senate but deleted by members of the joint congressional agricultural conference.

What resulted in the Tehuantepec isthmus was the Africanized Honeybee Cooperative Program, monitoring the movement of the bees through Tehuantepec by means of strings of trap hives, both on the Atlantic and on the Pacific coasts.

The plan used in Lost Hills, California, in 1985 was devised by Jeffrey Stibick of the USDA Animal and Plant Health Inspection Service. Stibick had also developed a plan for the Gulf Coast, and another for the thirteen-hundred-mile Texas border, which would, he estimated, cost $170 million.

You might suspect that a man who plans for insect emergencies would have emotional leanings toward the tragic, that he would have certain career attachments to the catastrophic. Perhaps this was true of Stibick, and perhaps not.

Jeffrey Stibick said that when the Africanized bee entered the United States, there would be "a severe decline in the pollination of many kinds of crops resulting in enormous crop losses. But we can't say that publicly," he said. "We can only say that there will be losses of a magnitude we cannot define. The only thing we can definitely say is that the almond crop in California will certainly suffer a lot."

There would be many stingings, and thousands of hospitalizations.

"The thing that bothers me the most," Stibick told me, "has to do with the quality of life. If you can't open your doors and let your children play in the yard because there are bees nearby, or if the children are playing in the backyard and you have to tell them to come running if there are any bees around, what do you have? Or if there is a grandmother or grandfather out in a chair, what about them? These are the people who were stung in South America, but normally an aware person who has his wits about him is not going to have any problem getting away. You might get stung a bit, that's all.

"When Africanized honeybees get in an area, there is an initial distress among humans and animals for a while, and then that declines as they learn to live with the bees.

"I've got a plan on the shelf to stop the bees along the Texas border, but it will cost the taxpayers an awful lot of money. It will be worth it, however. We have to have as good a plan as we can to wipe out these bees," Stibick said. "We can save the bee business for that much longer, you see what I mean?"

The Texas Action Plan, to be administered by the Texas Department of Agriculture, will impose quarantines within a 150-mile radius of any find of Africanized bees. Every hive inside the circle will be sampled. Initially, there will be depopulations, but once the Africanized bee is firmly established, depopulations will cease and a Management Plan will take effect. All hives within an Africanized zone will have to be requeened yearly. No unmarked queens will be permitted. Beekeepers will have to maintain 10 percent of their colonies for drone production, so as to saturate the area with European stock. All wild colonies will be destroyed.

Total economic cost, it is estimated, will be $83 million to $132 million a year for the first five years. It breaks down like this: $3 million to $4 million in beekeeping losses, $40 million to $80 million in agricultural losses from inadequate pollination, and $40 million to $48 million in lost tourism revenues. It is expected that these figures will decrease, however, as the bee population stabilizes, the public becomes educated, and management techniques are developed.

37

It was nearing sunset in Coleharbor. Jim Tiffany was standing on I-83, in front of the 83 Cafe, looking through binoculars. Chris English, another of the Bakersfield beekeepers, was with him. About a half mile up the highway at the top of a slope were two trucks, parked at the edge of a field.

Jim Tiffany lowered the binoculars and said, "Joe Romance is up there. He's got those guys from Florida. Could be trouble." Joe Romance—also one of the Bakersfield group—was not someone to fool around with.

"I'm going up there," Chris English said, in a way that meant he didn't want to miss anything, and he walked to his car. Jim Tiffany wanted to stay and look through the binoculars; he didn't want to get too close. I got into Chris English's car. It was a modified purple Gremlin with a racing stripe, a stick shift, and no reverse gear—a car he was very fond of—and we drove up the hill, stopping ten yards behind Joe Romance's flatbed.

The sun was low, and the sky was turning red. We were next to a freshly cut wheat field. Though the late light made the earth look black, the short stalks of wheat had a candlelike glow. Crickets were slowly marching across the section road.

The Florida truck was a faded blue and sat heavily on its axles. Two men, both looking very tired, were on either side of the truck, doors open, each with a leg inside—just in case. Romance stood about fifteen feet away from them, far enough so that they could hear but not close enough to intimate any sense of camaraderie. Tall, with a red beard, raised shoulders, arms away from his body, Romance looked ready for action.

"You can get mad at me," the Florida manager said compliantly. "We're never coming back here again."

Two weeks before, Romance had spoken with the manager, when he was looking for locations. The manager wanted to know if there was any space in Coleharbor. His area had dried up. Romance said there was room for everyone, that plenty of sunflower fields had not been covered. Romance knew that when sunflower flows, the fields can never be fully covered.

But then the beekeeper had found Romance's locations, and set his bees in Romance's yards, next to Romance's hives. One night, at 2 A.M., after loading bees since dark, Romance had driven into his yard and found bees there already, and in the best spot. Then Romance saw that the hives were from Florida, which meant they might have tracheal mites. Now Romance wanted to plow the hives over a hill, "The steam," he told me, "started to roll out of my ears."

"Are you going to Florida?" Joe Romance now asked the manager, meaning, you have a place to go.

"In a week," he replied.

"That's fine." His pitch rose slightly. "But I might not be leaving. I might not be *able* to leave. None of us from California will be able to go back." The manager nodded, apologetically, but in a way that said there was nothing he could do.

Just then a pickup truck came roaring up the section road and slid to a stop next to Chris English. The driver, a local farmer named Kranz, was having his troubles, too. Kranz had just left two other Florida beekeepers, who had placed a large number of hives on Kranz's land without asking his permission. Kranz had threatened to plow the hives over with a tractor if they weren't gone by morning, and one of the Florida beekeepers had replied that he would protect his hives with a shotgun. Kranz had started to get out of his truck, to fight, but Fred Tiffany stopped him, told him to leave, and promised he'd settle the matter.

"You see what I mean?" Romance said.

The manager nodded. "I know what you mean," he said.

But then something happened with Joe Romance. Somewhere during the exchange he relaxed and let his anger slip away. Judy Carlson had told him that this Florida manager was a good beekeeper, on top of things—except for this year—and after a while, after enough steam had rolled out of his ears, Romance had seen in the manager a picture of himself: a tired beekeeper with a common problem. Romance had looked at this beekeeper and he had seen another man.

"A human being," Joe said later, "and I'm standing there calling him a jerk." Romance told him to get out of town as soon as possible, then got into his truck, and drove home.

Joe Romance was the most ambitious and probably the most successful of the beekeepers in Coleharbor. He was twenty-nine, and he had already paid off a commercial loan of $100,000. In the former Coleharbor schoolhouse Romance stored honey for the federal support program, and he had an extraction plant, where he did custom extraction for other beekeepers as well. And despite the problems in American beekeeping, he wanted more hives.

Romance first got into beekeeping when he was a high school student in Los Angeles, a member of Future Farmers of America. He was in a math class, sitting in the back of the room, waiting for it to be over so that he could go surfing, when the teacher started talking about beekeeping. The teacher said that one hive could make fifty pounds of honey. Romance knew enough math to multiply fifty pounds of honey by $1 a pound and then to multiply $50 by one thousand hives. Soon he set up two hives. He played with them for two seasons and then lost interest.

One day Romance was backpacking in the Sierra Nevada with a friend, and the friend stepped on a rattlesnake. They hiked down out of the mountains, stuck their thumbs out, and waited for a ride. At 3 A.M. a flatbed truck pulled over and the driver told the hikers to climb on back. There was plenty of room—the truck was only partly loaded with hives. Romance and his friend had bees crawling over them, but Romance didn't mind. It took him back to high school. The driver was Jim Mayfield.

Romance went to work for Mayfield, and in 1977 they migrated to Coleharbor. Word was that North Dakota was the pot of honey at the end of the rainbow. There were no other beekeepers in Coleharbor, and they took locations wherever they could find them.

Joe received hives in payment for work. In 1977 he had thirteen hives, and by 1979 he had 450. In 1980, with a thousand hives, Romance went out on his own. Beekeeping was under his skin. In California, as soon as the grass turned brown it was time to get out of the state. It was just like birds going north.

We were at Joe Romance's house, sitting around his kitchen table. Chris English was there, and Joe's wife—Dawn Romance, a name for a beekeeper's wife if ever there was one. She was as beautiful as her name. They had a three-year-old son, who already had a hive of his own.

Dawn Romance wanted to know if I'd talked to other beekeepers' wives, and if they had also found it difficult to cope with the moving and the nocturnal hours. "I've been trying to go to college," Dawn said. "I want something for me."

"The beekeeping business is yours," Joe answered.

"I know the business is ours," Dawn said, "but it's not really my choice for a career. It's hard for me to get something going because we move around. I go to school a semester at a time, whenever I can. I went to summer school here, just trying to get it in. I don't have much patience," she said. "I have to keep reminding myself that it will happen.

"What's really interesting is the shocked expressions you get when people ask what your husband does and what you do. They go, 'A bookkeeper?' 'No, a beekeeper.' Everybody goes, 'Oooo.'"

In through the door came Jim Mayfield and Jim Tiffany. Fred had gone home to bed.

"I've got to decide what to do this winter," Joe said. "If I get into California, I could be quarantined. Maybe I'll have to start over with new bees in Texas. The economics of the thing is not all that different. If I kill my bees, I get fifty extra pounds of honey. I'll lose almond pollination but that'll save me six to twelve thousand dollars in truck rental. We won't have to buy queens and we won't have to move bees every night. Won't have to get uptight every day, coming home at night to pesticide calls, always saying hold the dinner, honey, I'll be back later.

"And they got the killer bees out there now in Bakersfield," Joe continued. "It's quarantined. It's a dying of the business. It's slowly dying right now."

"It's like the Ice Age," Chris English said. Among the Bakersfield group, and there were no what you might call straight arrows among them, Chris was known as the hippie (he'd been to Be-Ins), the educated eccentric (he'd taught art at a university), and the one good for the peculiar thought.

Joe didn't pick up Chris's theme, but responded anyway. "All farmers are in the same position," he said.

"Getting the dinosaurs out," Chris English said. "Metaphorically I think of these big guys, they are the dinosaurs, you see. Joe, Fred, Mayfield, all these operators."

"But I've been lucky," Joe said. "These other guys have had disasters. Truck rollovers, yards burning up, floods, you name it."

"Locoweed, gumdrops, red-hots," Chris said.

"We got locoweed in California," Joe explained. "When the bees feed on it, it makes them go crazy. They fly in circles. They become totally unproductive."

Chris explained, too. "This beekeeper fed his bees entirely on reject candy," he said. "Gumdrops and red-hots and candy canes. Got it from a factory for three cents a pound. He crunched the candy up in a big thing like a cement mixer, like a rock crusher. It would go around and around, and the dust came out and it hung on the ceiling and made icicles. It covered this guy, too. He looked like a snowman. He threw all the candy into a big pot and melted it and then gave it to the bees."

"They got sick," Joe said. "The bees all ended up dying on him."

"You'd find those little orange candy peanuts in the brood chambers," Chris said. "You'd find nuts stuck in the comb, marshmallows, everything."

Dawn burst into laughter. "Would you say beekeepers are generally eccentric?" she asked. "The few I know, it's like, what are they going to come up with next? I say to my family, these people are crazy. Don't worry about it."

38

"I got into beekeeping by accident," Chris English said. We were sitting in the 83 Cafe. "I was down at the bar one night, playing pool, and I happened to meet Joe Romance. He said, 'Buddy, can you paint?' I started painting hive covers.

"Initially Joe wanted to sell me bees, trade for work, and I said no. It's like adopting an orphanage. You're not going to take on a whole bunch of kids unless you want to play parent. And I realized that parenting is a part of beekeeping. So I waited, and I started off with eight hives. I made two hundred twenty-five pounds a hive. It was a good year."

Chris English was not someone who had made beekeeping his profession. He was a wanderer, someone who'd done many things. He had studied art, and had taught painting at the college level. He had shown his work in galleries in Seattle and then abandoned the painter's life. He had lived near volcanoes in Hawaii, in a cave in Arizona, under arched boulders in Guatemala (leaving when soldiers came looking for the jazz musician who was keeping the villagers awake), and had foraged for obsidian in California. He had even been a motel clerk. While a caretaker on a ranch, he met Joe Romance.

English was already fascinated with the Owens Valley, and beekeeping offered a new perspective on the floral life there. There were pollens in flame oranges, yellows and reds, rainbow pollens.

"Our cycle begins in September on the rabbit brush in the Owens Valley. The Owens Valley is about seventy miles north of the Mojave Desert, depending on how you measure. It is the deepest valley in North America, and from the southern part Mount Whitney is visible. Mount Whitney is the tallest mountain in the continental United States. So you have the lowest point and the highest point in America within fifty miles. Sometimes it's a real good pollen flow. There's always something, depending on the weather conditions.

"That's where the bees go dormant until early February, when we check them over. Then we move about a hundred fifty miles, to almonds in the Bakersfield area. The weather is different. Owens Valley is arid and cool. In Bakersfield it's warm and there's moisture. The bees respond in a dramatic way. They really come alive. It's like they wake up suddenly, and they're roaring on almonds.

"Oranges start to do something around Easter. I love the scent of oranges. I get overwhelmed by the bloom. The orange groves are just an endless canopy of white. There's a lot of lemon groves around, a lot of kiwis, and walnuts. There's apples now, too, Granny Smith apples coming in.

"We go back to the Owens Valley and get a little hit off the desert flowers. Depending on the conditions, you get different types of flowers in the spring. If it's a real wet winter, you'll get flowers that may not have come up for years. In early May we head up to North Dakota."

Chris was a sideliner. He had eighty hives of his own, and he didn't want a large operation. He talked of selling his hives. "I've seen the Fred and Joe movie, and I don't want to do that one. I've seen the range war here, and I don't want to get involved in that. I'm looking for a niche."

"Where is that?"

"An island someplace, with a twelve-month nectar flow. Clear water, exotic fruits and colors. The big island of Hawaii appeals to me because there are actually living volcanoes on it. I find that really thrilling."

Fred Tiffany, who had come into the 83 Cafe, was listening to Chris. "Have you ever thought of seeing a head shrinker?"

Chris English didn't reply. "I slept a few miles from one of the volcanoes. I could feel magma moving under the earth. And then Joe was saying hurry up, you've got to come back and help me move the bees. I came back to move bees into almonds."

I left Coleharbor. In Minot I met Everett Kehm, who had been keeping bees since the midfifties, migrating from Texas to Nebraska and ultimately to North Dakota. With a voice like soft brushes he told of the plants of the Texas brush country—Indian blanket, mesquite, horsemint, dewberries, bluebell—and he told of wanting to extract his sunflower honey and get out of town before the government pulled any surprises.

In Bottineau, near the Canadian border, I met Bill Hurd, who told me he was the meanest son of a bitch who ever lived, but it wasn't true; I'd met lots meaner sons of bitches. Hurd was quite kind, and he loved beekeeping. He'd come up in beekeeping, traveling with his father and a load of honey from Florida to Ohio when he was five, during the Second World War. Hurd's father had been one of the first, if not the first, beekeepers to migrate by car to Florida, strapping hives to the roof and fenders of a Model T in 1924. Bill Hurd had sold honey for a nickel a pound ("I mean there aren't many of us around anymore"), he had bought an operation in Costa Rica to work the Africanized bee firsthand, he had killed off mite-infested hives and after coming up mite-positive in North Dakota had refused to do it again. "You don't hardly run me out," Hurd said. "I'm a survivor, a mean sort of person that always seems to s-u-r-v-i-v-e."

In Dunseith, in the Turtle Mountain Indian Reservation, I encountered the sad sight of the Prettyman family extracting what honey they could from fifteen hundred hives depopulated with cyanide powder. A father-and-son operation, the Prettymans had kept bees since the fifties. And though they had come up with mite infestations in Florida, LaVerne Prettyman had decided to travel to North Dakota without permits, because not to use his locations was to lose them: "If you lose, you might as well do so with optimum results," LaVerne said. He was filing suit against the state of North Dakota to regain his locations (and would succeed), and extracting honey to pay for transportation costs back to Florida, and trying not to become cynical. "Look on the bright side," he said. "We run bees year-round. This way we get a break from it."

In the town of Devil's Lake a sharecropper who managed ten thousand hives, three thousand of them Horace Bell's, showed me his warehouse where he was storing two million pounds of honey for the federal government, and he showed me a bottling plant, funded by federal loans, where he bottled government honey for the federal giveaway program. And though it all had an overtone of wastefulness, it wasn't really; Roger Bracken was employing a dozen people in Devil's Lake. He was trying to develop foreign markets. He was bottling honey destined for school lunch programs and nursing homes, honey in plastic jars with a United States label on them. It was a charitable undertaking, on the government's part, and strangely, the effect it had on me was to make me feel patriotic pride—it wasn't weaponry, destruction, deception; it was a positive check on the slate of deeds. Bracken gave me a bumper sticker, a pocket calendar, a Bracken visor cap, and though he did it uneasily—honest man—a jar of government honey.

In Bismarck I stopped at one of the extracting plants of Powers Apiaries, a third-generation beekeeping corporation that had made the largest honey crop in the United States, 3 million pounds, but that had also, through the use of the honey loan program, become the largest defaulter of honey loans—2.5 million pounds forfeited, which seemed to cast a shade of dejection on Howard Schwab, the plant manager. "Do you think this is a dying industry?" he asked. "Do you think we're making buggy whips?"

In southwestern North Dakota I met Francis Andress, who ran Grand River Honey Company, the largest sharecropper in western North Dakota. Andress managed, on fifty-fifty shares, twelve thousand hives from five companies in California and Florida. He had three hundred locations in South Dakota and North Dakota. Andress had made a honey crop of 1.4 million pounds in the previous year—a hundred-fifty-pound average per hive—but this year, because of a drought, Andress expected a ten-pound average. He had five thousand of Horace Bell's hives, and the colonies looked as weak as newly formed splits. And they had mite infestations; I watched the Andress crew load Bell's hives to be escorted, by state police, from North Dakota.

And in New Salem I met Tom Emde, who had also come up in a beekeeping family. Tom Emde's company was called Sweet Briar Honey Farms. He lived with his wife and son on a hilltop on the North Dakota shortgrass prairie. Their winter home was in Apopka, Florida. Emde ran the standard cycle—orange, gallberry, palmetto, clover, alfalfa, sunflower. His hive count, by commercial beekeeping standards, was small, only seven hundred hives, but because Emde had grown up in beekeeping, worked in operations with thousands of hives, he had the skills to make a good living from a small operation. Emde, in fact, came as close as anyone I'd seen to living the idyllic dream of beekeeping—a comfortable, ordered life, migrating with the bloom.

It was late afternoon, and the hills were turning dusky brown. The Emdes, who knock off all work every Friday at sundown, were in their living room. Tom Emde, a lively, amiable man, told some of his story.

"My father, Earl Emde, started in California in the early thirties. By the time he finished high school he had several hundred colonies of bees and a truck. He began migratory beekeeping, on very poor roads, going up to Oregon, and going into southern California. He'd go out into the desert, and move into the higher elevations in the mountains for different crops. He'd go up to Oregon and back, up and down the West Coast.

"He started with a Model T, loading by hand. It was quite an ordeal. Little trails over the mountains. It was an all-night procedure just to get over one little range.

"In the forties he drove out to Iowa and Nebraska. He heard that there was good country back in the Midwest. He just took his truck and put a couple hundred hives of bees on it and headed east, not knowing anything about where he was going.

"His plan was to go to Iowa. He drove to Sioux City, and around Council Bluffs. He went all the way around Iowa in a big circle, with the bees on the truck. There was plenty of room in the country there.

"He remembered that there was an area back in Nebraska along the Missouri River Valley, yellow sweet clover, big, tall bushy plant. So he went back to Sioux City and located there, and along the eastern border of Nebraska he found places to put bees. They did so good he went tearing back to California to get another load and came back. That was the beginning of long-range beekeeping in California.

"He set up there. All through the forties and early fifties he moved bees from California back to Nebraska. The flow was terrific in those years, and he branched into South Dakota a bit. Real good years. He had maybe a couple thousand hives, and he was moving them back and forth by hand on a little straight truck with a sixteen-foot bed.

"In the 1950s he had had enough of the traveling and we stayed in Nebraska. We stayed put for about five years. And then he learned that Florida was good. He was always looking for new territory. He bought an outfit there, and we started moving from Nebraska to Florida in 1956. He was still pushing for new frontier, and we moved up into South Dakota in 1960. The farming practices had changed and there wasn't as much clover around Sioux City.

"All three of us boys were raised on beekeeping. We used to ride in the back of the truck behind the cab, my brothers Dave and Mark and I.

The sleeper was a wooden box in front of the load of bees, down on the floor. There was a little grate, and if we wanted Dad, we'd holler out there and eventually he'd hear us screaming, stop the truck, and open the little door, and we'd get out. We didn't dare open the door when we were moving because we'd fall out on the highway. I have a little problem with claustrophobia. Maybe it started there. One time Dad was in the back and Mom drove through a ditch. The hives shoved the sleeper in the back and he couldn't get out of it until they pulled it all loose.

"Because of the soil bank program, when farmers were paid not to use land, there were lots of acres of clover and alfalfa. It was a beekeeper's paradise. We had the clover up here and in Florida we had the orange groves. It was really tough moving out of the orange groves. The colonies expanded and they were really strong. You took the honey off and crammed the bees down into two boxes. When you picked them up the bees were all over the boxes, and on the truck was a solid wall of bees. It was much more difficult than anything today. We were still loading them by hand. One guy on the ground picking them up and another guy on the truck stacking them. If they were extremely heavy, you'd have two guys on the ground.

"What changed our whole migratory operation was when my brother Dave and I bought a forklift in 1966. When you do something new, you do the craziest things. We built pallets and we loaded them with all the entrances on aisles on the inside so they would all be shaded. We built a water system on the truck, a little nozzle system, and when we traveled through a town in the middle of the day we'd turn on our sprinklers. We'd be going slow through this town with the water spraying. We took a movie of our forklift operation and went to the beekeeping convention and that was kind of the beginning of palletized bees, as far as I know.

"In 1966 we came up to North Dakota. From the South Dakota line for a hundred-mile stretch we had the pick of everything. It was wide open. There was wild clover and there was alfalfa and that's where we made the big crops. That was the year of the fantastic bloom. We had such snow drifts here, and then in the spring a big flood. Some clover lasted all summer out here. The average, I'm sure, was well over two hundred pounds.

"In 1969 I split up with my brother Dave and ran my own bees. We had started up five thousand colonies of bees with Mark. It was all in the family. But in 1969 I said, 'Man, this is killing me,' loading all the bees on the truck like Dad had done for thirty or forty years. I took about seven hundred colonies of bees and ran my own business on the side. That's what I've been doing ever since.

"Our business is kind of small now, but we've been doing the same systems that Dave and I were doing when we were much larger. There's much more time for the family, it's much more enjoyable. You don't make as many mistakes. You're much more efficient if you have a hand in things. We hire summer help, high school kids. We don't have the expense of full-time help year-round. In the process of being more efficient we do all the stuff when it's supposed to be done. Life has been much more enjoyable, and we've made more money that way."

39

Honeybees are able to communicate the specific locations of nectar sources by means of a symbolic language. In darkness, on a vertical comb, in a crowded hive, scout bees perform miniaturized reenactments of their flights. These vibratory episodes, read by follower bees—again, in the dark, with outstretched antennae—carry information about both distance and direction to the goal.

Our knowledge of the information contained in the honeybee dances came from a lifetime of research by the German zoologist Karl von Frisch and his students. Von Frisch won a Nobel Prize in 1973.

Although we know the honeybee language, and can in fact pinpoint locations and plot a landscape by watching hivemates dance on a comb, we don't know how it evolved, how many millions of years it took bees to develop and genetically encode the behavior for a symbolic flight, but we can see some signs among the other bees. For example, there is the communication behavior of *Apis florea*, the dwarf honeybee, which lives in

Asia. *Apis florea* builds combs in the open, hanging them from tree branches or rock outcrops.

At the top of its combs *Apis florea* builds a platform, a flat dance floor on which foragers make short runs in the direction of nectar sources. The length of the run is about two body lengths of the bee, and when it is completed, the bee circles back and repeats the run. After a second run, the bee circles again, but along the other half of a circle. So the straight run is actually a diameter of a circle, and it contains not only the message "this way" but also "how far."

When *Apis mellifera* adapted to nest in enclosed cavities, it lost the benefit of a horizontal dance platform. Since it could not make short runs directly at the goal any longer, it had to alter the signal for direction to the nectar source. It had to learn to read on the vertical plane, and in the dark. But although *mellifera* changed the direction cue, it retained the same method for communicating distance to the source.

When the source is less than one hundred meters from the hive, the bees communicate the fact that food is available, and close to the hive, by vibrating their abdomens and moving in small, alternating circles. Follower bees reading the round dance rush from the hive and fly in the immediate area until they locate the source.

Scouts often do their round dances in specific territories within the hive, to an audience that is familiar with the floral source. Often the bees that respond to a round dance have been waiting for the flow to begin, and the round dance merely announces commencement. Thus, a specific area of a comb populated by a specific congregation of bees will be dedicated to a specific species of plant. This flower constancy is beneficial to both the hive, for harvesting honey, and the plant, for pollination purposes.

When the scout bee returns from the field and enters the hive, it gives out nectar samples. Follower bees can solicit nectar samples by making a squeaking sound. This sound is at a pitch of three hundred to four hundred cycles per second—a few notes above middle C. The squeak of a house bee halts the incoming forager.

When the flowers are beyond one hundred meters from the hive, a straight run enters into the round dance, and the wagtail dance begins. This dance is a series of straight runs followed by alternating half-circles. During the run the honeybee vibrates its abdomen thirteen to fifteen times per second. It also beats its wings thirty times per second, emitting a pitch of two hundred fifty cycles per second (slightly lower than middle C).

It is the number of circuits within a given interval that communicates distance—the tempo of the dance, how long the bee stays with its straight run. The straight run is a simile of flight, the bee saying to other bees, "Like this." The longer the bee takes to cross the tiny circle it inscribes on the comb, the longer the flight to the flowers.

Accuracy of information varies among individual bees. Older bees tend to dance slower tempos, indicating longer flights. When a bloom begins, scout bees communicate many distances, but the discrepancies decrease as the days go by.

Temperature may affect the distance message, as it takes more energy to fly in cold weather than in warm. A strong headwind or a flight up a steep slope results in a slower dance tempo, for the same reason.

The information about distance is not based on visual cues or on the amount of time it takes to reach a floral source. Rather, the information is about the amount of energy, or honey, it will take to fly to the goal. The honey stomach, used to carry and store nectar, is also a fuel tank. From other house bees the forager takes the proper amount of honey for the flight. For a flight of five hundred meters the honey stomach contains 1.61 milligrams of honey. A flight of a thousand meters requires 2.2 milligrams of honey. A flight of fifteen hundred meters requires 4.13 milligrams of honey.

On the return flight the bee carries a full load, as much as 75 milligrams of raw nectar.

40

"I'm going to be on the road for three weeks," Andy Card said. He had been on the road for eight hours, leaving Bangor at four in the morning with a bundled-up log house on the trailer, now heading west through Massachusetts on Interstate 90 for Buffalo. "Out here, down to Florida, up here again, back to Florida," he said over the engine noise. "Going to the places we always go in Florida, along the St. Johns River."

It was the second week of October. The foliage had turned, the woods along I-90 in the Berkshires were flame reds, oranges, and yellows. Andy's destination, after a stop in Albany to drop off the log house, was the town of Gowanda, twenty miles south of Buffalo, where Wayne Carter was waiting. Andy had moved 1,440 hives, four truckloads, to the Buffalo region to get a crop of honey from goldenrod. This trip was not only to pick up the hives, but to find out, after several weeks, how well they had done, or as the beekeeper would say, if they had made anything.

The hazardous trip south would follow. "You can't go to South Carolina anymore," Andy said. "It's closed to migration now. I have to call North Carolina, tell them I'm coming through. Georgia, I'm supposed to have a permit to go through there. Now Al Carl supposedly sent the paperwork to Georgia, so I'm halfway covered, but they want you to have the permit in the truck, which I'm not going to have for a trip or two. I'm gonna go through at night and hope they don't stop me, quarantine the load or kill all the bees or something. I'm only going through the state for a hundred and twelve miles. If I was going through the long way I'd make damn sure I had the permits. I just hope the state police won't hassle me.

"What we're hoping is that these hives out here pan out. There were a few things I saw out there that made me think this was a good area. But we'll have to see how they are. I don't want to brag it up too much."

When Andy made his July trip to Buffalo, he rented a car and fanned out from the city to look for stands of goldenrod in the farmlands. The goldenrod stalks were only a foot high then and a month from flowering. Goldenrod, in the sunflower family, is a fast-moving field plant that grows two to five feet tall and produces dense flower clusters that turn to airy seed packs soon after pollination. It is a heavy nectar producer, and goldenrod honey—dark, heavy, easily crystallized—is a source of winter food for bees. Some beekeepers harvest their crop just before the goldenrod flow and leave what follows for the bees. In some regions, mostly northern areas of the country, goldenrod is the last major nectar flow of the year.

Andy settled on an area south of Buffalo. The soils were a heavy clay, good for nectar production, and along with goldenrod there were several varieties of clover and a nectar plant called bird's-foot trefoil. With Lake Erie a few miles to the west there would be plenty of moisture and less chance of early frosts. From a company that charted weather patterns Andy bought printouts of rainfall levels, and he knew there was a high level of precipitation south of Buffalo. He figured that storms came off Lake Erie and dumped a lot of snow on the towns of Gowanda and Springville, and he knew from experience the correlation between snowfall and honey.

Andy had coursed through the farm country, walked into the fields, kicked up soil, studied the plants. He stopped at farms and knocked on doors. He gave away honey. It took three days of scouting to find the locations he wanted, but it took a lot longer to move 1,440 hives. They were not all in place until after the first weekend in September, and by then goldenrod was well into bloom. In addition to the possibility of a lost honey crop, and lost time and money, there was a bit of pride at stake. Though Andy's father had consented to the New York move, he had done so unwillingly.

We stopped at a construction site, where a crane was waiting to lift off the bundled log home, and where construction workers waited to turn it into a shoe store. Then later, at a truck stop Andy bought $100 worth of fuel, just to top the tanks off. He'd fueled up in Bangor; by his reckoning the semi was getting 5.5 miles to the gallon. Twelve hours on the road, red-eyed, Andy crawled into the sleeper for a nap, and I camped out in the diner.

It was night when we reached Gowanda, where we found Wayne Carter waiting in a motel room, and we all moved over to the motel bar. I had last seen Wayne at Horace Bell's yard. He had since left Bell, in April, to go back to work for Merrimack Valley Apiaries. Though Wayne had learned a lot working for Horace Bell, he wouldn't go back. It was too hot in Florida, and the work was too hard. Wayne said his hands were still swollen from the bee knuckle he'd picked up working in Deland.

Wayne had spent the day gathering the hives from all the outyards and collecting them into four yards of 360 each. It had been a hot day. The bees were all over him. They had stung him through his veil. But there seemed to be good weight. Sometimes, he said, when he was driving the forklift, it had bottomed out, tipped to its nose from the load on the forks.

Andy slept in the truck that night, and I took the extra bed in Wayne's room. But the sleep was short. We were up at six, and after coffee from a convenience store we drove to two concrete plants so that Andy could make arrangements to rent a scale. He had to know how much weight the hives had taken on, and if he would be under the legal limit for highway driving—seventy-eight thousand pounds.

We headed out of town, passing by many fields with ripe pumpkins, and passing by a billboard that announced that Gowanda was the home to the World Pumpkin Federation, and would soon host the World Pumpkin Festival, with major cash prizes. We drove by a continuous spread of dairy farms, and by vineyards with trellises stretching back over the hills.

We opened a pasture gate, followed a dirt road through a gully, and drove to a clearing on a hilltop. Andy climbed down from the cab, surveyed the goldenrod, and the views across the valley—green pastures cut into a mat of flame orange—and then he looked at the hives.

He pulled a jar lid from a cover and peered in. The frame tops were covered with bees, but Andy could see that there was white wax on the frame tops. "Plugged out," he said. His was the satisfied voice of the harvest. Andy pulled another lid, looked inward, pushed the lid back into place. "There's seventy pounds on this one," Andy said. He squinted as bees flew at him. "This one here's plugged out, too."

Andy strode to the truck. "We just got lucky," he said. "Had nowhere to go but up." Dreams of honey ascended. "If they're all like this, I figure maybe we grossed thirty grand on it."

"Time to do battle," Wayne said, and he put a leg into his bee suit. But a bee did battle first, and though we'd been on the hill fifteen minutes without a sting, Wayne took one on the wrist while he zipped up. "Hey," he said. "I didn't do anything to you." He flicked out the stinger.

It was a good day to load hives. The temperature had dropped down into the forties. There was a strong wind. Most of the bees would be staying inside.

While Wayne prepared to run the forklift, Andy rolled a net onto the empty truck bed, and he fastened this bottom net down with spring clips. Then he pulled some hay from a plastic bag and filled his smoker. He walked down the row blowing smoke into the hive entrances, and even though the wind was strong, the smoke drifted into the hives, caught on the air currents within.

Wayne set one pallet on top of another and tried to lift the pair.

"You see that?" Andy shouted. "Wayne couldn't pick up two pallets. That's a good sign." Heavy hives, heavy harvest. The hay in his smoker burned fast, and there were no pine needles on the ground, so Andy stuffed a dried goldenrod flowerhead into his smoker, puffed the bellows, and smoked on.

"This was pure roll-of-the-dice luck, getting these spots out here, pure luck." After all the difficulties of this year, it was good to see Andy's excitement. Honey was the reason for beekeeping. The immense scale of Andy's operation did not diminish the basic pleasure of the honey harvest.

Wayne loaded quickly. In forty-five minutes he had the first two tiers on, four rows of ten pallets, 240 hives. It took him only another twenty minutes to set the last tier of twenty pallets in place. Andy raised Wayne to the top of the load on the forklift tines, and Wayne walked along the hive covers, pulling back the net. It fell like a drape over the sides of the truck. Together they rolled the ends of the nets and fastened them with spring clips. It formed a neat package. "Tom Charnock developed this system," Andy said. "If anybody ever needed a tight load, it was him." They tossed straps over the net, hooked them in, and pulled on the ratchets with pipe handles.

I looked closely at the net, and saw two bees. One was on the outside of the net, and the other was on the inside, and they were mouth to mouth. These two bees—probably not even hivemates—were passing nectar. Every drop counts.

What a minuscule drop it was, after all, 75 milligrams of liquid. But it was a matter of scale—75 milligrams amounted to 90 percent of a bee's body weight. Proportionally, a 200-pound man would have to carry—not to mention fly with—180 pounds of water to match the haul of the forager. Those little drops, that fraction of the crop, would be reduced by the house bees. They would stretch the drops between their jaws, exposing them to the air. They would hang the drops on a comb to dry, like sheets on a line. And to quicken evaporation, the bees would line up in columns and fan their wings, making air currents that moved at a rate of 250 feet a minute. And of course they'd add their own glandular secretions, making finished honey from the flower sap.

Andy started the truck. The engine made a deep, guttural sound, and black smoke blew from the exhaust pipe. We climbed the hill, and when we turned onto the road, the truck rocked slowly and uneasily "It's top-heavy," he said. The gears whined as we rolled down the hill. "We definitely don't want to go over any wooden bridges now," Andy said.

41

The honeybee advertising a floral source on an open platform can communicate the distance to a goal by acting out a flight, producing sustained wing vibrations and measured sound pulses. Tempo corresponds to distance. But the distance component, with its linear formula, is the less complicated part of the language. Direction of flight is also communicated, and whereas the honeybee dancing in the light on a horizontal platform indicates direction by making short runs in the line of the goal, the honeybee inside a hive, running on a vertical comb, vibrating in darkness, has to transpose the visual cue to a symbolic one.

Two sensory organs come together to make the language symbolic: the eye and the gravity receptors.

The compound eye of the honeybee is composed of packed, conical lenses called ommatidia. There are approximately sixty-three hundred ommatidia per eye. The ommatidia are cone shaped and have convex outer lenses, and each separate lens, tending at a slightly different angle, gives its separate reading of the angle of light. The bee perceives the world as multiplicities of light and, when it is flying, flickerings of light, whether it be the flickering of flower petals or the flickering light of the sun. Flying along, its thousands of conical lenses collectively deducing the angular position of the sun, the honeybee reads the sky, and can remember the cast of light. For the honeybee, sky and sun form a compass.

Many insects, the honeybee included, have geomenotactic orientation abilities. In other words, they can walk on a vertical surface at a constant angle, a straight line up a wall. To do this precisely, the honeybee is equipped with sensory devices called gravity receptors, which are simply opposing bristles of hair. There are six pairs of gravity receptors, two on the joint between head and thorax, and four on the joint between thorax and abdomen. When the honeybee walks vertically, the receptors press against each other and the bee feels where it stands in relation to gravity, to straight up and down.

The peculiar success of the honeybee was to link the eye's ability to read the angle of light with the gravity receptors' ability to read the angle of body position, and to form a language from it. The bee transposed the visual cue (the position of flight according to the position of the sun) into a gravitational cue (the geomenotactic walk on a vertical comb). By aligning vision with the geomenotactic sense, the honeybee made a language in which *gravity* is the symbol for the *sun.*

The basic transposition is this: A walk straight up on a comb, or straight against gravity, means a flight straight toward the sun.

The dancing bee describes the two-dimensional realm of flight along the earth's surface, using a third dimension of reference, the vertical position, or azimuth, of the sun, as sunlight plays upon the ommatidia.

The straight run that is directly upward on the comb tells the follower bees, "Fly straight at the sun for a certain distance." When flight angle deviates, the straight run deviates in a corresponding manner. For example, if the flight is 45 degrees to the right of the line between the hive and the sun, the honeybee does its straight run 45 degrees to the right of vertical. And if the flight is on a solar angle 45 degrees to the left of the line between hive and sun, the bee does its straight run 45 degrees to the left of vertical. If the flight is straight away from the line between hive and sun, the bee does a straight run 180 degrees from vertical, or straight down on the comb. Dancing into gravity symbolizes a flight away from the sun. And so on, around the circle.

Of course the sun moves across the sky. The honeybee must compensate for the change. It does this with an internal clock that gives the bee the ability to memorize the speed of the sun across the sky. The honeybee needs to see the sun for only a few hours to learn the velocity of the sun's path for the entire day. The bee combines its ability to memorize this velocity with its geomenotactic abilities, and without leaving the hive alters the position of the straight run according to the changing position of the sun, with the ease of the hands of a clock. As the sun moves clockwise, the straight run turns counterclockwise.

Karl von Frisch also discovered a link between profitability of the nectar source and the liveliness of the dance. Certain highly excited individuals return to the hive and relentlessly advertise their discoveries. They remain in the hive dancing through the day and into the night, altering the straight run to create a gravity symbol that refers to the sun's position on the other side of the earth—a position the bee has never seen.

And while the excited bee is doing these marathon dances, other bees transpose the nocturnal solar angle back to a flight to a stand of flowers.

42

Andy shifted the gears of the Mack quickly, and he shifted them often—every few feet, when we passed through the town of Springville. No one took a second glance at the truck, probably because the nets obscured from view what was inside. We bucked up a long hill, and Andy turned on the road to the concrete plant. He traveled down an incline by a mountain of sand and came to a stop on a platform scale. Andy turned off the engine and said, "Now we'll see. Beekeepers lie. Scales don't."

Wayne Carter pulled up in the International. Together we followed the scale attendant into a shed and stood in front of the dial. The attendant released a lever, and we all watched as the needle sprung past 60,000, and 70,000, coming to rest at 74,480 pounds.

The load was 3,520 pounds under the legal weight limit. Andy could drive on the interstates.

That settled, Andy did the kind of figuring beekeepers love to do—

honey math. The truck weighed 28,000 pounds. Therefore, the load weighed 46,480 pounds. Divide 360 hives into load weight: 118 pounds per hive. Subtract weight of pallets, bees, wood, wax: roughly 70 pounds per hive. By Andy's reckoning (and assuming this beekeeper wasn't lying), the harvest came to roughly 50 pounds of honey per hive. For the 360 hives on the truck that amounted to 18,000 pounds of honey. He had four loads. If the other hives had done as well, Andy Card's bees would have made 72,000 pounds, or 36 tons, of honey.

More honey math: It was the second week of October. The hives had come into Gowanda and Springville during the first week of September. So they'd been working the fields five to six weeks. Roughly speaking, the 1,440 colonies had collected 36 tons of honey in about the same number of days. Roughly speaking—and this honey math is all rough math—the MVA bees had made a ton of honey a day.

But the honey crop was only a hypothetical one, since the honey would remain on the hives, with the bees, as a source of winter feed. This was Andy's father's decision, and since Andy Card, Senior, still owned Merrimack Valley Apiaries and had the power of decision making, the honey would remain on the hives and go south for the winter. But Andy followed his father's orders with gritted teeth. He wanted to sell the honey. Part of him wanted the load to be over 78,000 pounds, so that he would be forced to return to Billerica and extract the crop.

We left the scale and drove to a restaurant. "By not extracting the honey," Andy said, "we're saving nine to twelve thousand dollars in expenses for corn syrup. But the honey's worth about forty thousand dollars, so we're spending forty thousand dollars to save nine. That don't figure with me. That honey ought to be extracted and the money put in the bank where it belongs. The wax alone is probably worth twenty-five hundred dollars."

The difference of opinion between elder and junior Card was that of the old and the new schools of beekeeping, the traditional and the modern approaches. Andy Card, Junior, had spent many hours with Horace

Bell, and he was of the Horace Bell school: Feed the bees corn syrup, not honey, especially if the bees are wintering in the south and don't need to manufacture heat. And furthermore, field bees are of no benefit after the harvest—they are just straws sucking honey, so take them off the hive with the honey supers.

But Andy Card, Senior, was old school. Feed the bees honey. Take strong and intact colonies into the winter, even if the large populations required greater honey supplies, even if those populations caused food shortages, and possibly starvation, in the early spring.

Andy would follow his father's orders. It was give and take. Andy had taken some risks earlier in the year, and his father had gone along. And now Andy would relent. He had, after all, succeeded in Gowanda. "Part of the reason my father is worried is because we're not going ahead, we're going backward. The profit margin is decreasing.

"I had hopes of building this thing up to five thousand hives, cranking out a half-million dollars a year, gross, and everybody can live pretty well on that. I just don't see it on the horizon. Too much cheap competition. Pollination prices are dropping, they're not going up. More beekeepers are getting into it, and some have been operating with one foot in bankruptcy court.

"Most of the people we compete with, they're like owner-operated trucks. You have a truck, and you make a living with the truck. You make payments to the bank, but chances are you don't make a return on your investment. Chances are you're just buying a way to make a salary. Generally the people in the bee business are not getting a return on their investment. They're getting a salary, and they're damn happy to have a salary. Someday they'll stop and say, is that all there is?

"You know, we're really not making much honey. And the prospects of making honey in Florida are diminishing. There'll be orange honey in Florida again, but not until Wesley and Glenn get older. But I don't want Wes and Glenn to follow in my footsteps. I think I'm going to steer them into a position that's in demand. Lawyer, doctor, business manager, engineer.

"I don't know if we'll continue to migrate, either. We're going to Florida this year, but we're going to have to come up with something different. It may be that we'll just wise up, take what we've learned in migrating, and winter over in New England, just do a better job of it. Maybe we'll save money in the long run."

Lunch was over, and it was time to part ways. Andy would head south through Pennsylvania.

We stood outside the restaurant. "Keep in touch," Andy said. "Next crop will be maple, December fifteen."

He climbed into the cab. The trailer was low, only a few inches above the wheels, and the mudflaps nearly reached the ground. The engine started and the cab shivered, but the load settled easily as it moved onto the highway. Semis were made for the big interstates, and the interstates for them. The bees had merely become a part of the constellation of movement.

Andy Card rolled south, into the next season.

43

If an obstacle such as a hill or building blocks the line of flight to a nectar source, the bee does not broadcast the angle of the detour. Rather, it communicates an angle of flight straight at the goal, through the obstacle, but lengthening the distance component, thus telling of the energy needed to fly the detour.

If there is a strong crosswind and the bee must fly at an oblique angle, it does not communicate the angle of its body in the crosswind. That would be an undependable message, since winds shift. Instead, the bee deduces the angle of flight as if there were no wind, and the follower bees, knowing the correct solar angle, make the necessary adjustments in flight.

On a mildly overcast day the bee, up to a certain point, can still see the sun, because it can perceive ultraviolet light, and ultraviolet light penetrates clouds. When it is too cloudy to see even the ultraviolet rays, the bee sometimes uses a bright spot in the sky as a substitute for the sun. In utilizing this "lamp effect," the bee adds a secondary symbol to its language—gravity as symbol for the bright spot, which is in turn a symbol for the sun.

If the sun is obscured but there is a patch of blue sky with a visual angle of at least ten to fifteen degrees, the honeybee reads and utilizes the pattern of polarized light in the sky to determine the sun's position. Humans are unable to see polarized light, but many other animals can. The vibration of polarized light roughly follows a pattern of concentric circles around the sun, with deviations at two neutral points, and it has the same diurnal course through the sky as the sun. The honeybee eye contains receptor cells that correspond to the polarization pattern and are most sensitive when the polarization is parallel to the axis of the receptor. Flying from the hive, the follower bee reads the polarization pattern in the blue patch of sky, deduces the position of the sun, and then aligns its flight according to the correct solar angle to the goal.

It's remarkable that the bee can do these things—broadcast distance by tempo, indicate direction by symbol, use substitutes for the sun, compensate for crosswinds, use polarized light to deduce the sun's position—but perhaps it is even more remarkable that the follower bee can read the signals, as it stands on the vertical comb in the dark, with outstretched antennae, among a group of other follower bees, and at an angle to the communicating bee. The follower reads the dance, transposes the gravitational angle to a solar angle, relates the dance tempo to distance, reads the polarized light to determine the sun's position, compensates for a crosswind, and flies to the flowers.

The accuracy of the followers in finding the goal is sometimes greater than the information in the dance. Language is symbolic and metaphorical, and by nature both elastic and inexact, so the follower bee does not read merely one flight reenactment. It follows several dances, and averages the results.

Dialects exist for the distance component among the various races of bees. Some strains dance slower tempos than others for the same distance traveled. The Italian bee is among the slowest of European dancers, and African races dance slower tempos than the Italian. The longer tempo would have a link to metabolism, one could assume. An African race such as *Apis mellifera adansonii* would consume more honey during the same mile-long flight than the Italian bee, *Apis mellifera ligustica.*

Von Frisch also discovered something he called misdirection cues. If the cover is removed from a hive, and the dancing bee has a direct view of the sun, it discards the gravity symbol and uses the sun's actual position as a reference point. *Apis mellifera* sometimes dances on the entrance board of a hive, on the horizontal plane, using the sun as reference, just as the tropical *Apis florea* does, reverting to a previous mode of language, of times before the enclosed nest.

If, however, the honeybee is advertising a nectar source and has a view of the sky but not a view of the sun, two points of reference come into competition. The bee has the option of using either gravity as a symbol for the sun—the walk straight up and down—or the actual position of the sun as deduced from the pattern of polarized light. Sun and symbol for the sun compete, and a confusion could result, but the bee bisects the two angles, and does its straight run according to the bisected angle. The follower bee actually sees no misdirection. It reads the bisected angle, computes the two competing angles—gravity, sun—in relation to the goal, and transposes them to the correct angle of flight.

The misdirection cues usually occur when bees swarm and are hanging in a cluster. The swarm cluster is a mass of bees consisting of an outer surface, an entryway, and an inner area with branched chains. After the swarm has settled on a tree branch, scouts leave to look for nesting sites. Potential sites are inspected for adequacy of size, and protection from moisture, temperature, and wind. Upon returning to the swarm the scouts dance on the surface of the cluster, advertising their find. Follower bees read the dances, leave to find the potential site, make their inspections, return to the cluster, and dance according to their preference.

At the beginning of the selection process many sites are usually available, and multiple factions form. Sometimes as many as twenty sites are advertised. However, the most suitable sites are advertised with greater enthusiasm, and increasing numbers of scouts convert to the better sites. Inferior sites are thus eliminated, and the factions reduce in number. Gradually a decision is made, and the best site is selected. Sometimes the selection takes a few hours, and sometimes it takes days. In some rare cases, no decision is made, and the colony is left to perish in the open.

Dances vary in their orientation on different parts of the cluster. Those bees with a view of the sun tend to dance using the sun as a reference. Those with a view of the blue sky dance the intentional misdirection cues, the bisected angle of two directions. But since followers are subject to the same influences, no confusion results.

One spring I watched a swarm come to the process of decision. It took several days. What was initially a loosely hanging, cylindrical cluster of bees altered its form as time passed. The mass of bees seemed to take on tension, to form intelligence. The loose cluster pulled up, and stretched horizontally, into the shape of an ax blade. It looked like it was going to pull apart. It looked like a difficult decision in process, a visible, palpable mind making itself up.

Scouts initiate the dispersal of the cluster by making loud zigzag buzzing runs. Cluster and queen are herded into the air by the scouts, which course through the cloud of bees and fly along the edges of the swarm, leading the way.

Once inside the new nest, the bees set about making comb. They have come with loads of honey from which to make wax. Before the cells are completely drawn, the queen begins depositing eggs. Scouts inspect the area, and begin advertising nectar sources, according to the gravity symbol.

There are eccentrics in the bee colony, in this world of precision and determined response. Some bees read the dances wrong, and search in all directions but the ones advertised. But they, too, are biologically important to the colony, because they find new floral sources. And although they read the dances wrong, apparently these eccentric, non-conformist bees communicate their discoveries correctly.

44

Four seasons passed. Horace Bell's colony count stayed between twenty thousand and thirty thousand beehives, and in 1986, with a honey crop of nearly three million pounds, he became one of the three top honey producers in the United States.

Reggie Wilbanks increased queen production and honey production to offset losses in the Canadian package-bee market. Tom Charnock left beekeeping. Dale Thompson and Jeff Kalmes married and left beekeeping. Jim Owens continued to broker beehives. Young Jimmy Owens left beekeeping, but hoped to return. Chris English sold his hives and moved to Washington. Fred Tiffany returned to the glass business for a year, and then returned to beekeeping. Joe Romance increased his colony count.

Growers in the Maine blueberry barrens increased hive rentals from ten thousand to thirty thousand hives, approximately sixty semi-loads of leased bees.

Andy Card increased his colony count to eight thousand hives. In addition to investing in a two-thousand-hive operation in Louisiana, Card also set up an operation with a breeder in Manitoba. Thus he had two sources of stock—one in the south for mite-free bees, and another in the north, a sure and safe supply in the event of quarantines after the arrival of the Africanized honeybee. In 1989 Card also moved a semi-load of five hundred hives from Louisiana to California and pollinated in almond orchards. Merrimack Valley Apiaries thus became, Card believed, the first commercial beekeeping company to pollinate in both the eastern and the western sectors of the United States, and with rental fees of half a million dollars per year, MVA was perhaps also the country's largest pollinator. On a new Peterbilt semi, Card stencilled a drawing of a cowboy riding a honeybee, and the words "Crop Pollination—Nationwide."

Two seasons after the tracheal mite crisis, the varroa mite appeared in the United States. Varroa, an external pest even more debilitating to colony life, brought on another crisis. More regulatory measures arose, primarily at the state level.

Varroa could have been treated with a pesticide called fluvalinate, but that set off another controversy. Many beekeepers, especially noncommercial beekeepers, wanted to avoid the use of pesticides inside their colonies, and understandably so—after all, the honey was inside the hive, and pesticide use would taint the natural purity of beekeeping. These people preferred to prevent the spread of the varroa mite—and thus the migration of beekeepers. The state of Connecticut banned the transport of bees on semis. Massachusetts, with a greater need for commercial pollination, came up with a ruling that all bees coming into the state must be certified as mite-free.

A recent estimate indicates that the number of bee colonies in the United States has decreased by 25 percent, from four million to three million. That trend will continue with an increase in winter losses from mites and with the arrival of the Africanized bee.

In North Dakota, Bill Hurd said that commercial beekeepers would deal with the Africanized bee. "I know if I had a hive of African bees, they'd be dead before sunup," Hurd said. It would be in the commercial beekeeper's interest.

The commercial beekeeper, like Hurd, would not only kill off a hive of Africanized bees, he would also replenish it with gentler stock. The hobbyist beekeeper, one of that other 99 percent of the beekeeping population, would also be likely to kill off the Africanized colony, but chances are that he would not soon replenish the hive. Chances are that the hobbyist beekeeper would quit beekeeping. The beekeeping population would thus decrease by one beekeeper and one or several hives. It's likely that as the Africanized bee moves through Texas and Louisiana, many beekeepers and their hives will drop off along the way. At least that's the way things happened in Latin America.

People like Hurd, Card, Emde, Wilbanks, and Bell are the ones who can respond to the spread of Africanized bees, and will help to curtail them. Not just because it's in their interest, but because they're the most able. And even if the commercial and migratory beekeepers' approach to bee husbandry can seem excessive, cold, and objectionable at times, their skills and their vision are capable and impressive.

"I don't do nothing but think about bees twenty-four hours a day," Horace Bell said to me, while he stared into his fireplace one January in Florida. Bell, and others like him—wizards, Andy Card called them—are the ones who can see into the situation and come up with answers. Nearly all these wizards need to move from one region to another. As there are rattlesnake roundups in Reggie Wilbanks's hometown of Claxton, Georgia, there will later be killer bee roundups. Such localized efforts will probably follow the first, large-scale bureaucratic responses.

At the end of the summer of 1990 the Africanized bee had not quite reached the Texas border. A freeze in the previous year had killed off forage in Mexico, and a drought in 1990 slowed regrowth and the migration of the Africanized bee. But swarms were appearing in the trap lines near Soto La Marina, just 150 miles south of Brownsville.

And then, on October 15, it happened. A swarm of Africanized bees was caught in a trap in Texas, in the Lower Rio Grande Valley, east of the border city of Hidalgo. The bees were destroyed, a quarantine was imposed, and a new era in American beekeeping thus began.

As trek swarms bypass traps and move along lowlying areas, following promising flora, it will be important not to strike at beekeepers—hobbyist, sideliner, commercial, or migrator—for whatever ultradefensive behavior these new bees may exhibit. For such behavior is rooted deep in the land, and won't be the fault of those who keep bees.

As for me, I now have one hive. It sits behind a shed, and the bees stream through the branches of a cherry tree into the air. I hadn't opened it for several years, but this season I did. I cleaned the frames off, I'll work it through the season, and maybe, for the first time in a while, I'll take some honey off.

I like to watch them, most of all, and now, when I see them making their sweeping arcs, when they glide down among the crowds of bees at the hive entrance, I just watch. Contraction has followed expansion, and I sometimes think of a zen saying: At first mountains were mountains, and then the mountains were not mountains, but some other thing; now mountains are mountains again.

That sentence comes closest to describing where I am now. I'm not professing enlightenment, or special knowledge, or anything like that. Only that for me, in some way, bees are bees again.

Epilogue

Billerica, June 2003

Though Andy Card had said in 1985 that he wouldn't encourage his sons to take up the beekeeping profession, that turned out not to be the case. He actually did leave the door open for Wesley and Glenn.

On a day in early summer at Greenwood Farm, eighteen years since I'd last seen him, Wesley was building a toolbox for one of the flatbed trucks. Using a new power saw, he cut the pieces for the frame, and these pieces he welded together. Wesley was determined to get the job done that day, because he had to get on to pouring a concrete floor in one of the barns, so they could use it for honey and wax storage.

Wesley and Glenn had both recently returned from the Maine blueberry barrens, where there had been a quality inspection of the hives, a "midterm exam" as Wesley called it. The result of the exam was important. In 2003 the price of a hive rental had increased to $50, or $300 per pallet, but blueberry growers in Maine gave beekeepers a bonus for high-quality hives that could bump the rental fee closer to $60. Merrimack Valley Apiaries was renting 4,000 hives in the barrens. In a few weeks they would move a portion of them to Cape Cod cranberry bogs.

In May, Wesley had graduated from Cornell University with a degree in agricultural business management. Glenn had just finished his freshman year at the University of Vermont. He intends to study agriculture and business, and to join the beekeeping business in another three years. Now, while Wesley built the toolbox, Glenn loaded a pickup truck with cakes of beeswax to deliver to a client who used them to make candles.

Both boys seem to have Andy's boundless energy. Wesley was brimming with ideas. In college he had utilized MVA as a research subject for many of his courses. In a computer-modeling course he had studied transportation costs. During his senior year he made a study of the cranberry industry. For a marketing course he wrote a proposal for a fictional business called "Crystal's Pure Honey Company," offering unprocessed, natural honey, with the intention of capturing some of the health food market in eastern Massachusetts. His report stated that per capita consumption of honey had increased by 35 percent from 1990 to 2000, according to government figures; that Chinese imports had been banned in 2002 due to the presence of chlorophenicol; that Argentine exports had decreased with the economic collapse in that country. All the more opportunity for marketing a health-oriented product, Wesley had argued.

Greenwood Farm no longer served as the holding yard for beehives for MVA. Many more houses have been built along the road, including one next to what used to be the holding yard for MVA bees, and where semis could be unloaded. Now at Greenwood Farm there is a little Christmas store where they sell wreaths and trees grown here and in New York.

"We don't keep bees here anymore," Wesley said, but then added, "except for those few down there." He gestured to a field beyond the house, where in a pocket of spruce trees about fifty hives stood. For a hobbyist

Glenn and Wesley Card after delivering hives to a cranberry bog on Cape Cod. —GLENN CARD

beekeeper fifty hives would be a demanding part-time business, but by MVA standards fifty hives could be considered an insignificant amount, representing about one-half of 1 percent of the colony count, now pushing towards 10,000. But Wesley didn't consider the colonies at Greenwood insignificant. What he'd meant was that the neighborhood was too developed for them to bring in semi-trucks loaded with bees.

A swarm was taking to the air—a unified vibratory episode—swirling up like a golden apparition when Wesley and I walked down to the small apiary. As we stood near a plastic cubelike container for high-fructose syrup, Wesley said that MVA buys eight to ten semi-loads per year now for feeding the bees. They had just fed colonies in Maine, and would do another feeding when the bees were on cranberries.

Andy had given both Wesley and Glenn the responsibility for moving hives to Cape Cod and for communicating with growers. It seemed a wise decision on his part to let the boys develop the cranberry accounts. There had been a crucial shakeup in the industry in the 1990s, and the field was ripe for development.

Following cranberry pollination, they would transport 3,500 hives to western New York to the farm south of Buffalo. Andy was there now. In September Wes would go to New York to help extract the goldenrod crop. That would take about six weeks. In January, for the first time, he would travel to the Louisiana farm to split hives and become more familiar with that operation.

"I was ready to do it full-time three years ago," Wes said. "I can't wait to be involved in all aspects of it, and contribute."

While the boys did their jobs, Crystal worked in the office in the farmhouse. When Glenn had started college last year, Crystal began raising queens again. She had spent the past winter in Louisiana raising queens for the nucleus hives that MVA was now selling to hobbyist beekeepers in Massachusetts. The nucs were made up in Louisiana and trucked to Billerica in the spring. The Cards had been in the nuc market for two years now, selling 150 in the first year. They sold 450 in 2003 after Crystal joined the operation.

"I'm going to take the nuc business over, make it my baby," she said. She felt free now that the boys were able to take care of themselves, and she loved Louisiana. The pace of life was slower there, she said, and the people were less materialistic and more friendly. She intended to go there every year and eventually raise queens for all the MVA hives, to be used in the yearly requeening process.

"With honey prices up, we can do more and reinvest in the company. We've got new blood in the business now, which will take the pressure off of Andy. I can administer and do the nucs. Andy can do the beekeeping thing."

Crystal smiled and said, "Andy can think like a bee."

Otto, New York, July 2003

Andy owns a former dairy farm in the town of Otto, fifty miles south of Buffalo. An old silo is still standing on the grounds. He has converted the barn to a storage building, woodworking shop, and wax-melting room. Andy also built a warehouse for honey extraction. It's a typically large-scale commercial setup, with mechanized uncapping lines, four large extractors capable of holding 168 frames each, and huge settling and storage tanks. With this unit Andy's help is able to extract 2,000 pounds of honey in a single run. They can extract up to 14,000 pounds per day. In 2002 they extracted 250,000 pounds of honey at this farm.

On a hill beyond the warehouses in Otto, set away from the road, Andy has built a house where he stays for much of the year. He hired Amish carpenters who hauled logs from the woods and fashioned cabinetry of black walnut and beech. To one side of the house are four ponds, which Andy has rejuvenated and stocked with fish. There is a field of Christmas trees, and a field of silage corn—an exchange with a dairy farmer for bee locations.

On the wall in his office are county maps, with colored pushpins denoting the locations of bee yards. Andy also keeps soil maps, to track where to set hives. His hive locations—because of higher yields—are almost all on areas of high limestone soil content. The pins show bee yards in areas around Otto, but also in another region to the east.

The maps and pushpins give his office the look of an ongoing military campaign. (Perhaps it's no coincidence that Andy's bookshelves are packed with military histories.) It's not difficult to see connections between the territorial acquisitions of armies and the scouting, plotting, and acquisitions of the commercial beekeeper. Generals aren't the only ones dealing with field forces numbering in the thousands.

Aware of this strategizing, now a CEO with operations in New York, Louisiana, California, and New England, Andy jocularly calls himself "El Presidente Grande." He's grown older and his sideburns are graying, and as for the physical labor required, Andy is now, at fifty-two, semi-retired. He walks with a limp, the result of a jump off a loaded truck after a long night spent retrieving hives from apple orchards. Andy now tells his employees not to get too tired. From all the years of hard work lifting hives and heavy supers, he has developed osteoarthritis in his lower back. As a result, he oversees more now. He thinks like a bee.

These days Andy hires trucks, rather than running the interstates himself. The last semi-rig he purchased sits in a garage, unused. He has fourteen employees—ten in Louisiana, two in Otto, two in Billerica (Wesley and Glenn).

One morning before the two New York employees arrived, Andy talked about the developments in his business. He owned the business now—he had made peace with his father, and told me he was grateful to him for starting Merrimack Valley Apiaries fifty years ago.

"After my dad turned the business over to me in 1986 I bought a brand-new flatbed truck, and that winter I bought the outfit in Louisiana. Horace Bell had told me about Louisiana. Probably the most significant event in our growth was when I bought eight hundred hives from Bob Hayes. He has an outstanding knowledge of bees. He had worked for Overby Apiaries, at one time the largest queen and package producer in the U.S. To my surprise, when I bought the operation Bob offered to manage the hives. He said that we could split the hives and increase the operation, and also manage for honey production. Over the next ten years we increased to two thousand five hundred hives, and each year made from five hundred to one thousand five hundred hives to send north. Under Bob we increased our honey production to two hundred fifty thousand pounds.

"When Bob retired in 1997 I bought a hundred-acre ranch in Bunkie, Louisiana, and we installed Joe Sanroma as our new manager. Joe had worked for us in Billerica since he was fourteen. I call Joe my third son. It was rough for him at first, but he's developed into one of the best bee managers around.

"Everything was going along great, everything clicking. We had three thousand hives rented in apples, six thousand five hundred hives in blueberries, and six thousand five hundred hives in cranberries. Good crops in Louisiana. Honey prices had improved. Then in 1997 we suffered what I call the 'cranberry crash' in Massachusetts, when the market got flooded with cranberries. The growers were getting eighty dollars a barrel when the market peaked, and in a little over a year the price dropped to ten dollars. Our pollination contracts on cranberries dropped from six thousand five hundred to two thousand five hundred. That was two hundred thousand right off the top.

"In the old days, if we lost a single pollination contract I had a heart attack right there on the spot. We lost so much business in cranberry pollination after the crash that I was numb, and it was out of our control. Honey prices dropped too. The last time I was on orange I sold honey for ninety-five cents a pound, on the hives. But by the time the cranberry crash struck, it had gone to fifty cents for orange and forty-two cents for Louisiana tallow.

"We were struggling and it made me realize that we had to diversify. We had to be less dependent on pollination. One thing I did was to start looking for markets for beeswax. We had been hoarding wax, stacking it up. I thought it was worth more than what they were paying. After all, you only get fifty pounds of wax per ton of honey. Wax was a good way of diversifying, and wax production increased with honey production. We're selling sixty thousand dollars worth of wax this year, from nothing. It's a good step in the right direction.

"About two years ago we saw the chance to sell nucs to the hobbyist beekeepers in Massachusetts. We offer a five-frame nuc with a laying queen, and as of now we're experimenting with a guarantee. The nuc business is something I really enjoy. People come up and get their box of bees and cradle it like a newborn child. Some of them even have a name for their queen. And you know they're going down the road and looking at what's in bloom. I'd hate to go through life without having those thoughts."

Andy was thinking of marketing pollen, too—the perfect food, he said, increasingly sought out by health food enthusiasts. With Wesley and then Glenn coming into the business, there was one other crucial area to explore, the direct marketing of honey to the consumer—the idea as expressed in Wesley's report on the Crystal Pure Honey Company.

Last, there was the California operation, managed under a shared arrangement by Lee Cashman, a beekeeper Andy met when he was pollinating almonds. "As we speak," Andy said, "Lee is moving bees to irrigated cotton and alfalfa locations for honey production."

In 1985 when pollination was the primary enterprise, Merrimack Valley Apiaries produced 100 barrels or 66,000 pounds of honey. In 2002 they made 1,000 barrels, approximately 660,000 pounds for a tenfold increase. That year, due to the ban on Chinese imports and the reduction of imports from Argentina, honey prices leaped. Andy had been getting 42 to 46 cents a pound, but in 2002 the price leaped to $1.40.

"We bumped up close to a million-dollar crop. But as far as honey prices go, we're in the NFL. That stands for 'Not For Long.' With all agricultural products, it's always 'not for long.'"

Even with all the new developments and the effort to diversify, pollination has remained a mainstay of the business, the foundation it was built upon. The blueberry markets in Maine have become especially strong, with growers now renting three to five hives per acre (up from the one and a half per acre in 1985), and beekeepers now transporting 50,000 hives into the barrens. Apple accounts, however, have decreased, by as much as 25 percent, with a drop in apple prices and the rising values of New England real estate.

As for the cranberry accounts, though Andy thought Massachusetts growers would never recover from the 1997 crash, there were some signs of an increase. "We went from two thousand seven hundred to three thousand two hundred rentals this year. Another trailerload more."

Andy made his calls that morning—to Wesley about shipping hives from the Cape, to a honey buyer, and to a trucking firm ("We'll be shipping a load a week starting in October," he said when he placed his order). After lunch, he and an employee, Russell Michener, went to a dairy farm and loaded a defunct milk tank onto their truck to use as a honey storage tank.

Andy then drove to his "experimental yard" in Gowanda. Gowanda is a fairly large town, with a business district and busy thoroughfare. Andy drove through a neighborhood and up a hill to a high bluff that overlooked the town and the Zoar Valley beyond. He drove along a grassy corridor between two lines of houses, where off near a garage, and by a line of trees near the edge of the bluff, were about twenty-five hives.

The man who owned this land was very interested in bees, and had encouraged Andy to bring 50 or 100 hives to this place. Andy, worried about the proximity of the spot to the neighborhood, thought it would be best to start smaller. The hives had done extraordinarily well here, averaging about 200 pounds per colony. Some had recently been requeened with Russian stock, considered to be more resistant to mites.

Beekeepers have continued to cope with the effects of mite damage, which includes severe winter losses if hives go untreated. Mites have rapidly acquired resistance to a limited arsenal of chemicals used to control them, and according to Andy Card, a "super mite" now exists that is resistant to all chemicals. The ultimate solution to mite infestations may be in the genetic approach, through the use of races such as the Russian honeybee Andy has been using to requeen some of his colonies, and through the development of so-called "hygienic bees," selected for traits such as cleaning mite-infested larvae out of brood comb.

Andy intended to overwinter hives here in Gowanda, and he had been considering eventually overwintering as many as 1,000 colonies in western New York. He had been thinking that overwintering would take the stress off the bees, if they were freed from migration. Honey crops might be higher. Transportation costs could be spared.

"The biggest experiment of all is survival," Andy said

And so it seemed that the migratory beekeeper, in his efforts to diversify, would be moving his bees less extensively. He called these colonies, the ones that would be staying in place, "indigenous hives." It seemed to Andy that, like his dealings with hobbyist beekeepers in the nuc business, overwintering hives would be a return to the kind of beekeeping he had done with his father years ago. "We're coming full circle again," he said. "Those are cool things to be doing."

He fired up his smoker, put on a veil and hat over his T-shirt (a T-shirt that read MVA BEE PUNCHERS, with a drawing of a cowboy riding a bucking bee), and went to work checking the hives.

The hobbyist beekeeper, using new boxes with three coats of paint, might look askance at the hives Andy kept on the bluff in Gowanda. They were shaky-looking stacks, with flaking paint, weathered wood, and more entrances for the bees than might seem proper. But they were tall, really tall, with a half-dozen supers, many of them, or more—you needed a lot of supers to get 200 pounds if you were extracting all at once.

He pulled off a cover and looked down into the frames. "This one's plugged out," he said, before going to a stack of supers, getting one, and adding it to the hive. He put the cover on, somewhat loosely—this hobbyist, the one watching him, went over and tried to fit it on more tightly, to close the openings, but to no avail.

Andy opened another, and said, half jokingly, "I told the queen supplier that the boss would be watching those Russians so they better send their best stock." He got another super, smacked it on the hive, and replaced the cover. He checked another. It was plugged up too.

"That's good, huh?"

He went through several more hives, stacking on supers, taking no stings.

On the way out he said, "Can you guess why this is such a great location? I may have to tell you."

"Basswood? Suburban landscaping?"

"It's not the plants."

"High on a hill?"

"You're getting close. It's air drainage. Cold air flows down that hill. This is a great spot for overwintering."

He drove down though Gowanda again, but before he left town Andy made a diversion, to a cemetery. He wanted to show me the gravestone of a beekeeper.

Someone had shown the grave to Andy a few years ago, but now he had a little trouble finding the site. He drove and scanned the graves, looking for the one that would be so familiar. Round he went, down a hill, up again, saying that it couldn't be in the old part, that the beekeeper hadn't died all that long ago.

Then, when he drove into a cul-de-sac, he said, "Right there," and stopped the car. And there it was—a granite headstone in the shape of a hive with three deep supers. Honeybees were etched into the sides. The beekeeper's name was on the top super, and read, "Edward Gabel, 1908–1981." His wife had passed away a few years ago, and her name was on the second box. The name of a daughter, still living, was on the lower box above the entrance board.

What a delightful idea that seemed, a Langstroth headstone. There was nothing ostentatious about it, nothing that seemed to be intended to call attention to itself. Architecturally it blended in perfectly with its surroundings. It seemed appropriate in what it suggested—peace, a touch of originality, a life among the bees.

"He was the beekeeper who was here before I arrived," Andy said. "Everywhere I went I kept hearing his name."

"Commercial beekeeper?"

"He made a living at it. Five hundred to eight hundred hives."

Andy left Gowanda and sped across the countryside then. It was true farm country, field after field, pastures as far as the eye could see, mile after mile. Many of the farms had that abandoned look, with disused dairy barns and empty silos.

Andy wasn't noticing that, however. Moving across this landscape, he talked about the floral geography.

"See that, that's spotted knapweed over there."

"Where, where?" I said. I'd not seen spotted knapweed before, or not seen it as a honey plant.

"There's a stand of basswood trees down in that valley," he said.

Basswoods, I thought—my favorite tree, with its white clusters of flowers hanging like bells in early summer, its beautiful fragrance, and honey of supreme quality. A regal tree, basswood.

Andy looked down toward the bank of a river. "Those are tulip poplars down there." The bloom had gone by, but the feeling, looking at them, was of anticipation.

Goldenrod was everywhere, the young plants without blooms yet. Hives would soon be on their way from New England to meet that flow. Andy loved this countryside. Driving through it, speaking of his occupation, he said, "Look at where we get to work!"

He showed more enthusiasm as we neared Buffalo, and passed by a railroad yard and manufacturing complex. This was an area of high limestone soil, Andy knew, and in a wide swath along the rail bed was a tall stand of white sweet clover, a premier honey plant if there ever was one. Andy happened to be talking about shipments of queen cells by mail then, but interrupted himself and pointed to the snowy bloom.

"Look at that!" he said. And he began to wonder how he could get there.

I keep three hives now. I've found that for me it's the right number. I can increase from one or two in the event of winter losses, and at least one of the hives will usually produce a substantial honey crop. I use Buckfast bees, in honor of Brother Adam, the esteemed British monk and beekeeper who developed this gentle, productive and mite-resistant strain at Buckfast Abbey, and who wrote about the world's races of bees.

Brother Adam also was an authority on mead, the fermented drink made of honey. His mead was aged in oak casks, and he claimed that it had to be aged at least five years to come to maturity. It seems such a promising investment, to convert honey to mead, put it away for years, and wait for the results. With crops of more honey than I can use or give away, I have found that mead is a delightful way to utilize that honey. Depending on whether herbs or fruits are added, mead can vary from a clear and sparkling beverage to something as deeply colored as port wine. I've made mead with additives of rose petals (Rose Petal Mead—a fine name), apple cider (a drink called cyser), grapes (a drink called pyment), blueberries, blackberries, peaches, vanilla bean, ginger root, and even blends of tea. I have a mead aging composed of honey and maple sap.

But honey and mead are not the only rewards of the effort. I keep bees mostly so that I can watch them. I like to see them fly home in the light of the setting sun. I like to stand near the hives and listen to their collective hum. I stand nearby in the draft of nectars pushed by those wings. I enjoy tracking their paths—over the hayfield next door, down the hill to a dairy farm with field clovers, to an old foundation site shaded with locust trees, to the blackberry stands, and so on. Bees are consciousness-raising creatures, in that they extend human consciousness to the floral landscape. You might say that this is a product of the angiosperm evolution too.

I have learned that the fieldwork of Karl Von Frisch, the Nobel Prize–winning scientist who discovered bee languages, has been duplicated by an American biologist, but without the same results. A possible conclusion arises that honeybees do not do the wagtail dance, or do not use gravity as a symbol for the sun. What a disappointing prospect, if such findings are true—and they may or may not be true.

But you can't doubt the honeybee intelligence. You only have to observe them to see it. For example, there is a birdhouse thirty feet from my three beehives, and for the past two springs tree swallows have taken up and nested there. Trees swallows are impressive creatures, too, as they soar and dip over the fields hunting insects, and as they pair to parent their young, one dropping out of the entrance to the house as another glides in to feed the nestlings. They have purple backs, long wings, and delicate feet.

A honeybee has to be an irresistible meal to an insect-catching bird, and I've often seen the swallows dip by the hives, especially early in the season. Somehow, however, word of this gets back into the hive, because I have also seen, overhead, a honeybee or even a trail of three or four, chasing after the swallows. It seems remarkable that the bees chase after the birds, but not me, that I can stand a few feet under this pursuit and yet not be the object of it. They know who the enemy is, and somehow they've been able to broadcast it.

Winter will come. I'll bank the hives against the wind and snow. I'll dig them out in the winter after storms, and listen to their zizz as they come to the entrances. On sunny days, some will fly out, and I'll find them on the snow. I'll think ahead, to the melting, the maple buds, the dandelions. I will wonder what form their anticipation takes, how they will ready for the bloom, those unified minds, contracting into their compact furnaces only to expand that intelligence to the fields around them.

Notes

For much of the information about the movement of beehives worldwide I drew upon a book by Brother Adam, *In Search of the Best Strains of Bees*, Northern Bee Books, Dadant & Sons, Hamilton, Illinois.

For much of the history of migratory beekeeping and of the package bee industry I drew upon Frank Pellet, *History of American Beekeeping*, Collegiate Press, Ames, Iowa, 1938.

For information about honeybee behavior and insect communication I relied upon Edward Wilson's book *The Insect Societies*, The Belknap Press of Harvard University Press, Cambridge, Massachusetts, 1971.

For information about honeybee communication I relied upon Karl von Frisch's work, *The Dance Language and Orientation of Bees*, The Belknap Press of Harvard University Press, Cambridge, Massachusetts, 1967.

For an explanation of the history of the honey support programs I relied upon USDA economists Harry Sullivan and Jane Phillips.

For information on pollinating behavior I relied upon *The Hive and the Honeybee*, Dadant & Sons, Hamilton, Illinois, 1976.

For information about pollination activity in the Maine blueberry barrens I received information from Dr. Amr Ismail, of the University of Maine, Orono.

For information about activities during the Africanized bee incident in Lost Hills, California, I relied upon the "Africanized Bee News," Division of Plant Industry, California Department of Food and Agriculture, Volume I, Nos. 1–4, 1985.

Much of my information on the Africanized honeybee of Africanization came by way of Dr. Thomas Rinderer, of the USDA honeybee research labs.

Much of my information about events during the honeybee tracheal mite crisis and other events came from the monthly news report of the *American Bee Journal.*

OTHER SOURCES OF INFORMATION WERE AS FOLLOWS:

Anderson, Earl D., *An Appraisal of the Beekeeping Industry*, Washington, D.C., Agricultural Research Service, U.S. Department of Agriculture, 1969.

American Bee Journal, Hamilton, Illinois, "$8 Million Africanized Bee Barrier Proposed," November 1986.

Bailey, L., "Reflections on the Discovery of *Acarapis woodi* in the United States," *American Bee Journal*, Hamilton, Illinois, February 1985.

Benson, D. Keith, "Africanized Bees: Their Tactics of Conquest," *American Bee Journal*, Hamilton, Illinois, June 1985.

Benson, Keith, "American and Canadian Beekeepers Study Africanized Bees in Venezuela," *American Bee Journal*, Hamilton, Illinois, March 1985.

Caron, Dewey M., "Entomological Society Holds Africanized Bee Symposia," *American Bee Journal*, Hamilton, Illinois, March 1985.

Cobey, Susan, and Timothy Lawrence, "Status of the Africanized Bee Find in California," *American Bee Journal*, Hamilton, Illinois, September 1985.

Cobey, Susan, and Sarah Locke, "The Africanized Bee: A Tour of Central America," *American Bee Journal*, Hamilton, Illinois, June 1986.

Collins, Anita M., and Thomas E. Rinderer, "The Defense Behavior of the Africanized Bee," *American Bee Journal*, Hamilton, Illinois, September 1986.

Congressional Record—House, October 7, 1985, H8331–8334.

Crane, Eva, *The Archaeology of Beekeeping*, Cornell University Press, Ithaca, New York, 1983.

———, *Honey, A Comprehensive Survey*, Heinemann, London, 1975.

Dadant, C.P., *Langstroth on the Hive & Honey Bee, American Bee Journal*, Hamilton, Illinois, 1922.

Danka, Robert G., and Thomas E. Rinderer, "Africanized Bees and Pollination," *American Bee Journal*, Hamilton, Illinois, October 1986.

DeJong, David, Roger A. Morse, and George C. Eickwort, "Mite Pests of Honey Bees," *Annual Review of Entomology*, 1982.

Eckert, J.E., and F.R. Shaw, *Beekeeping*, Macmillan Company, New York, 1960.

Federal Register, Part III, Department of Agriculture, Animal and Plant Health Inspection Service, "Subpart—Honey Bee Tracheal Mite; Revocation; Final Rule," April 18, 1985.

Free, John B., *Insect Pollination of Crops*, Academic Press, 111 Fifth Avenue, New York, 1970.

Frisch, Karl von, *Bees: Their Vision, Chemical Senses and Language*, Cornell University Press, Ithaca, New York, 1971.

———, *The Dance Language and Orientation of Bees*, The Belknap Press of Harvard University Press, Cambridge, Massachusetts, 1967.

———, *The Dancing Bees*, Harcourt, Brace & World, New York, 1953.

Goltz, Larry, "Migratory Beekeeping," *American Bee Journal*, Hamilton, Illinois, January 1987.

Government Accounting Office, by the Comptroller General, Report to the Congress of the United States, "Federal Price Support for Honey Should Be Phased Out," August 19, 1985.

Hamilton, W.D., "Altruism and Related Phenomena, Mainly in the Social Insects," *Annual Review of Ecological Systems*, 193–232.

Heinrich, Bernd, *Bumblebee Economics*, Harvard University Press, Cambridge, Massachusetts, 1979.

Killion, Gene, "Move Your Colonies with Care," *American Bee Journal*, Hamilton, Illinois, 1982

Laidlaw, Harry H., *Contemporary Queen Rearing*, Dadant & Sons, Hamilton, Illinois, 1979.

Levin, M.D., "Honey Bees Do Pay Their Way," *American Bee Journal*, Hamilton, Illinois, May 1986.

——, "Value of Bee Pollination to United States Agriculture," *American Bee Journal*, Hamilton, Illinois, March 1984.

Lindauer, Martin, *Communication Among Social Bees*, Harvard University Press, Cambridge, Massachusetts, 1971.

Martin, E.C., and S.E. McGregor, "Changing Trends in Insect Pollination of Commercial Crops," *Annual Review of Entomology*, Volume 18, 207–26.

McDowell, Robert, "The Africanized Honey Bee in the United States: What Will Happen to the U.S. Beekeeping Industry?" U.S. Dept. of Agriculture, Economic Research Service, Agricultural Economic Report 519, November 1984.

McGregor, S.E., *Insect Pollination of Cultivated Crop Plants*, Agricultural Research Service, United States Department of Agriculture, Washington, D.C., 1976.

Miller, C.C., *Fifty Years Among the Bees*, A.I. Root, Medina, Ohio, 1911.

Morse, Roger A., and Richard Nowogrodzki, "Trends in American Beekeeping, 1850–1981," *American Bee Journal*, Hamilton, Illinois, May 1983.

Moffett, Joseph O., *Some Beekeepers and Associates,* Joseph O. Moffett, Route 3, Box 175A, Cushing, Oklahoma, 1979.

Pellet, Frank C., *American Honey Plants,* Dadant & Sons, Hamilton, Illinois, n.d.

Rinderer, Thomas E., "Africanized Bees: An Overview," *American Bee Journal,* Hamilton, Illinois, February 1986.

Roberts, Radclyffe B., "The Evolution of the Honey Bee," *American Bee Journal,* Hamilton, Illinois, November 1975.

Robinson, Gene E., "The Dance Language of the Honey Bee: Controversy and Its Resolution," *American Bee Journal,* Hamilton, Illinois, March 1986.

Root, A.I., E.R., H.H., and J.A., *ABC and XYZ of Bee Culture,* 39th Edition, The A.I. Root Company, Medina, Ohio, 1983.

Severson, D.W., "Swarming Behavior of the Honey Bee," *American Bee Journal,* Hamilton, Illinois, March 1984.

Stricker, Milton H., "Pollination Pioneer," *Gleanings in Bee Culture,* Medina, Ohio, October 1971.

Swezey, Sean L., "Africanized Honey Bees Arrive in Nicaragua," *American Bee Journal,* Hamilton, Illinois, April 1986.

Taber, Steve, "Bee Behavior," *American Bee Journal,* Hamilton, Illinois, October 1984.

United States Department of Agriculture, Agriculture Handbook Number 335, "Beekeeping in the United States," 1980.

United States Department of Agriculture, Agricultural Research Service, "An Analysis of Beekeeping Costs and Returns," Production Research Report, No. 151, 1973.

United States Department of Agriculture, Economic Research Service, Agriculture Information Bulletin Number 465, "Honey: Background for 1985 Farm Legislation," September 1984.

United States Department of Agriculture, Economic Research Service, Agriculture Information Bulletin Number 497, The Food Security

Act of 1985, "Major Provisions Affecting Commodities," January 1986.

United States International Trade Commission, Honey—Report to the President. USITC Publication 781, June 1976.

Voorhies, Edwin C., Frank E. Todd, and J.K. Galbraith, "Economic Aspects of the Bee Industry," Bulletin 555, September 1933, University of California, Berkeley, California.

Washington County Regional Planning Commission, "Washington County Facts," Machias, Maine, 1981.

Wilson, Edward O., *On Human Nature*, Harvard University Press, Cambridge, Massachusetts, 1978.

About the Author

Douglas Whynott has taught writing at Columbia University's School of the Arts and the University of Massachusetts. He currently directs the MFA program at Emerson College. Whynott's books *Giant Bluefin* and *A Unit of Water, A Unit of Time* were both critically acclaimed works of narrative nonfiction. His next book, a nonfiction account of the lives of three country veterinarians, will be published in 2004. He lives near Hanover, New Hampshire, where he tends his own beehives.